GENDERING PLACE AND AFFECT

Attachment, Disruption and Belonging

Edited by
Alex Simpson, Ruth Simpson
and Darren T. Baker

BRISTOL
UNIVERSITY
PRESS

First published in Great Britain in 2024 by

Bristol University Press
University of Bristol
1–9 Old Park Hill
Bristol
BS2 8BB
UK
t: +44 (0)117 374 6645
e: bup-info@bristol.ac.uk

Details of international sales and distribution partners are available at bristoluniversitypress.co.uk

British Library Cataloguing in Publication Data
A catalogue record for this book is available from the British Library

ISBN 978-1-5292-3275-2 hardcover
ISBN 978-1-5292-3276-9 ePub
ISBN 978-1-5292-3277-6 ePdf

Cover design: Liam Roberts Design
Front cover image: unsplash
Bristol University Press uses environmentally responsible print partners.
Printed and bound in Great Britain by CPI Group (UK) Ltd, Croydon, CR0 4YY

FSC
www.fsc.org
MIX
Paper | Supporting responsible forestry
FSC® C013604

Contents

List of Figures

Notes on Contributors

Darren T. Baker, PhD, is Assistant Professor of Responsible Leadership at Monash University, Australia. His work provides an ethical and psychoanalytic analysis of the key challenges and opportunities in leadership, business and society, including sustainability, corporate social responsibility and diversity. Follow him @Darren_T_Baker.

Rajeshwari Chennangodu is Assistant Professor in the Organizational Behaviour and Human Resources area at the Indian Institute of Management, Kozhikode. Her broad research interest is in the sociology of work and ethnographic research methods. She is specifically interested in understanding alternative ways of organizing work.

Dan Harris (they/them) is Research Professor in the School of Education, RMIT University, Melbourne, Australia, and Co-Director of Creative Agency research lab: www.creativeresearchhub.com. Harris is editor of the book series Creativity, Education and the Arts (Palgrave Macmillan) and has authored, co-authored or edited 22 books and over 150 articles and chapters as well as plays, films and spoken word performances. They are activated in their scholarly work by creativity, affect and queer theoretics, and are committed to the power of collaborative creative practice to drive social change.

Anna Hickey-Moody is Professor of Intersectional Humanities at Maynooth University in Ireland. She is the chief investigator of a large research project into the educational experiences of queer religious youth and has written about youth subcultures for two decades. Her most recent book is *Faith Stories*, published in 2023 by Manchester University Press.

Nicholas Hill is McKenzie Postdoctoral Fellow in the School of Social and Political Sciences at the University of Melbourne. He is the author of the edited volume *Critical Happiness Studies* (Routledge). His research focuses on LGBTIQA+ (lesbian, gay, bisexual, transgender, intersex, queer/questioning, asexual) experiences of mental health and suicidality. He specializes in community-engaged research and using innovative participatory methodologies to produce practical resources for service users, community organizations and government.

Alison Hirst is Associate Professor of Organization Studies at Anglia Ruskin University, UK. She has published research on the role of spaces, places, technologies and other material objects in organizing. She is currently working on a European project that investigates the impacts of advanced manufacturing technologies on human embodied skills. Alison has published articles in *Organization Studies*, *Organization*, *Gender, Work and Organization* and *Public Administration*.

Jessica Horne is Lecturer in Management at the University of Sussex. Her forthcoming doctoral research uses creative qualitative methods to explore the social and affective dimensions of volunteering in house museums connected to the Bloomsbury Group.

Troy Innocent is an artist gamemaker, urban play scholar and creator of 64 Ways of Being, an augmented reality (AR) art trail platform blending game design with live and public art. He is Director of the Future Play Lab at RMIT University, Melbourne, Australia, working across speculative design, creative place making and urban play to explore new ways of being in the world. Innocent develops AR games that blend physical objects with digital interfaces to reimagine everyday urban environments in playful ways in public spaces from Barcelona to Hong Kong, Melbourne to Taipei, Ōtautahi to Singapore. Working with the city as a material, his 'reworlding' practice explores posthuman methods that reimagine, reconfigure and reconnect with the world.

Katherine Johnson is Professor and Dean of the School of Global, Urban and Social Studies, RMIT University, Melbourne, Australia. Her interdisciplinary and international research aims to improve the lives of LGBTQ+ people, particularly in the areas of queer youth mental health, transgender access to health, LGBTQ+ palliative care and LGBTQ+ suicide prevention. She is the author of *Sexuality: A Psychosocial Manifesto* (2014) and Routledge Series Editor Transforming LGBT Lives (with Kath Browne). Her research has been funded by National Institute of Health Research (UK), EU Horizon 2020, National Health and Medical Research Council (Australia), Suicide Prevention Australia, the Australian Research Council and VicHealth.

George Kandathil is Associate Professor in the Organisational Behaviour area at the Indian Institute of Management, Ahmedabad. He has two broad research interests. The first is related to power dynamics, particularly domination and reaction to it, within change in organizing work and workers. Specifically, he explores issues on new technologies, gender, organizational space and culture. His second broad interest is in subaltern alternatives to corporate and monopoly capitalist modes of organizing work and workers. Specific interests are in worker-owned cooperatives, ecofeminist alternatives, work solidarity-based alternatives in informal space and commoning-based alternatives in informal space. He has published research articles related to these topics in scholarly journals such as *Economic and Industrial Democracy*, *Gender, Work & Organization*, *Journal of Business Ethics* and *Organization*. He has also published a research monograph on transformation of family-owned business into worker cooperatives through employee takeover.

Evgeniia Kuziner is a junior researcher of the Centre for Youth Studies and a PhD student at the Higher School of Economic, St Petersburg, Russia. Her main research interests include gender studies, homelessness and gender aspects of homelessness. From 2018 to 2022 she conducted qualitative research into women's experience of homelessness in St Petersburg.

Murray Lee is Professor in Criminology at the University of Sydney Law School. His research interests focus broadly on representations of crime, media and processes of criminalization. He is the author of over 70 scholarly journal articles, book chapters and books, including *Inventing Fear of Crime: Criminology and the Politics of Anxiety*.

Patricia Lewis is Professor of Management, specializing in gender and entrepreneurship and gender in leadership, in the Kent Business School, University of Kent and is a fellow of the British Academy of Management. Her current research draws on the concept of postfeminism to critically investigate the gendered aspects of entrepreneurship and leadership. She has published widely in a range of journals including *British Journal of Management*, *Gender, Work & Organization*, *Gender in Management: An International Journal*, *Human Relations*, *Organization* and *Organization Studies*. She was Joint Editor-in-Chief of the journal *Gender, Work & Organization* from mid-2017 to the end of 2020 and she has recently been appointed as editor of the *International Small Business Journal*.

Paul McGuinness is Lecturer in Criminology and Sociology at the University of Sussex, where he is a co-director of Sussex's Crime Research Centre. His work seeks to combine criminology, hauntology and arts-based research. Most recently, he is the co-author, with Alex Simpson, of 'Hauntological cinema: Resisting epistemic erasure and temporal slippage with Sorry to Bother You', *The Sociological Review*, 70(1): 178–198.

Rachel Morgan is Lecturer in Human Resource Management and Organisational Behaviour at Brunel University London. Her research broadly investigates the impact of inequality on the experiences of marginalized groups at work. She has co-authored book chapters in *The Sage Handbook of Qualitative Research Methods* (2018) and *Stigmas in the Organizational Environment* (2018) She has co-authored articles in *Gender in Management: An International Journal* (2020), *Population, Place and Space* (2021) and *Sociology* (2022) Her current research focuses on structural inequalities and challenges to identity and subjectivity of working-class males in dirty occupations.

Nyk Robertson (they/them) is the Senior Diversity Officer and Assistant Dean for the Institute of Interdisciplinary Studies at Emporia State University in Emporia, Kansas. They are a member of Chi Alpha Epsilon and Phi Kappa Phi honours societies. Dr Robertson recently completed their dissertation titled 'Dialogue and Hope in the Classroom: A Poetic Inquiry into Freire's Theories Effect on Student Activism'. They enter this work through a lens that centres belonging, humanity and a pedagogy of hope.

Nick Rumens is Professor in Business and Management at Oxford Brookes University, UK. His main research interests include lesbian, gay, bisexual, trans, queer sexualities and genders and organizations, queer theory, management and organization studies, workplace friendships and organizational masculinities. His research has been published in peer-reviewed academic journals including *Human Relations, Organization Studies, Academy of Management Learning and Education, Sociology, International Small Business Journal* and *Gender, Work & Organization*.

Christina Schwabenland is Reader in Organizational Behaviour at the University of Bedfordshire and Director of the Centre for Leadership Innovation. She has published extensively on non-profit organizing, including an anthology on *Women's Emancipation and Civil Society Organisations* (Policy Press, 2016). Her recent work has concentrated on the implementation of equality, diversity and inclusion policies within higher education and she is currently the holder of three grants exploring these issues through the medium of LEGO® Serious Play®.

Corina Sheerin is Lecturer in Finance at the National College of Ireland, Dublin. Corina is also a Chartered Financial Analyst. Her research agenda is underpinned by a feminist stance and is multidisciplinary in nature. Situated at the nexus of gender, work and organizations, Corina has published widely in a range of Chartered Association of Business Schools peer-reviewed journals. Her work has been recognized internationally, having been awarded the Monica Lee Award for Outstanding Research in 2021 from the Academy of Human Resource Development, the Emerald Best Paper Award and a Highly Commended Award in 2019 for both the *Journal of European Training and Development* and *Gender in Management: An International Journal* and the 2016 Alan Moon Prize at the 17th International Conference on Human Resource Development and Practice. Her current research is focused on issues of gender and space, gender performativity and gendered organizations.

Alex Simpson is Senior Lecturer in Criminology at Macquarie University. His work draws predominantly on ethnographic research methods to examine the production and maintenance of elite cultural spaces. This has led to a critical focus on the 'performative rituals' of elite social groups, the maintenance and development of financial landscapes and distinct masculinist identities that sustain neoliberal 'traits' of competition, aggression and risk seeking. He is author of *Harm Production and the Moral Dislocation of Finance in the City of London: An Ethnography* (2021) and has published in leading journals, including *Sociology, Cities, Sociological Review, Gender, Work & Organization* and *Ethnography*.

Ruth Simpson is Emeritus Professor at Brunel Business School, UK. Her research interests include gender and careers, masculinity in non-traditional occupations and gendered dynamics in 'dirty work'. Her recent project, funded by the Leverhulme Trust, has explored the experiences of male manual workers in UK seaside towns. She has published widely in international journals including *Human Relations, Sociology, Work, Employment & Society* and *Gender Work & Organization*.

Melissa Tyler is Professor of Work and Organization Studies at the University of Essex. She is co-director of the Essex-based Centre for Work, Organization and Society, and of the Future of Creative Work group and co-chair of the Gender, Work and Organization Judith Butler research network. Her research on gender, sexuality, feminist theory, ethics and the body has been published in a range of authored books, edited collections and journal articles. Melissa's recent books include *Judith Butler and Organization Theory* (Routledge, 2019) and *Soho at Work: Pleasure and Place in Contemporary London* (Cambridge, 2020). Her current research (with Philip Hancock) focuses on lived experiences of precarity and recognition among freelance workers in the creative industries.

Introduction

Alex Simpson and Ruth Simpson

To be in space is to be affected by it. As we move through space, we become wrapped within the material, social and technological banalities of everyday life, framed by quickly forgotten encounters, repetitive commutes and moments of disruption. Passing through our placed environments, be it within the city, suburban or rural frame, is, so often, about the mundane; a pre-reflexive coding of anticipated and experienced social encounters, which is enveloped within an ongoing relational interplay with the cultural, material and 'atmospheric' world that surrounds us. It is here that, drawing on Amin and Thrift (2002: 30), we can see place's spatial ecologies framed by variable events – twists and fluxes of interrelation – rather than specific moments of encounter, fixed in space and time. In other words, place can be seen as a 'constellation of processes' (Massey, 2005) that comprise and incorporate a distinctive and multiple range of differentiated spatial relationships (Halford and Leonard, 2006: 669). Within this flux, we experience the dynamic process of encounter, re-encounter and sense making, the pre-reflexive feeling that gives rise to senses of belonging, community, security as well as ripples of fear, exclusion or dis-ease. The ordering of the social and material framing of space, from aesthetics to mobilities, soundscapes to architecture, can – and will – elicit a felt response within each of us. This 'poetic representation', to use Young's (2023: 210) words, of physical and geographical space can be brought together to the framing of *affect*.

In what has become a canonical text, Massumi (2002) compares affect, which he describes as 'a resonation, or interference pattern', to an echo:

> An echo, for example, cannot occur without a distance between surfaces for the sounds to bounce from. But the resonation is not on the walls. It is in the emptiness between them. It fills the emptiness with its complex patterning. That patterning is not at a distance from itself. It is immediately its own event. [...] With the body, the 'walls' are the sensory surfaces. The intensity is experience. The emptiness or in-betweenness filled by experience is the incorporeal dimension of

the body. [...] The conversion of surface distance into intensity is also the conversion of the materiality of the body into an *event*. [...] This is not yet a subject. But it may well be the conditions of emergence of a subject: an incipient subjectivity. (Massumi, 2002: 14; original emphasis)

The focus is on the echo, containing atmospheric, material and sonic reverberations, that resonates *in between* the 'walls' of our surrounds and within ourselves to capture the felt force of affect. It frames how, fundamentally, affect privileges the body as the site of intensity. Here we can see how affect is founded on the emotive experiences and force-relations that exist in between, and that run through, body to body, human and non-human (Gregg and Seigworth, 2010). For Matthews and Wall (2023), affects are deeply felt – in the stomach, on the skin, on the tip of one's tongue – before they can be fully articulated, explained or narrated. The prism of an echo creates an imaginary where affect is seen as a process of encounter – be it soundwaves encountering the different textures of surface, or our own pre-reflexive encounters with the material and social world through which we move. As Conradson and Latham (2007) argue, as we move through the social and material world, across placed ecologies, we each interact in a multitude of ways through the realm of material, imaginative and social encounters. To this end, affect has been described by Amin and Thrift (2002) as a form of thinking; a whole-body cognition of understanding, experiencing and responding to the world.

Stepping off from these insights, this book contributes to the framing of place and affect by revealing the ways the background ordering of social life shapes – and is shaped by – our pre-reflexive and unarticulated anxieties, hopes, fears, loves and hates, which lay latent within our experience of the social and material world. In other words, as a central argument running through this edited collection, to be in place is an *affective* process. After all, since affect emerges as a relation between bodies, objects and technologies, it has distinctly spatial characteristics (Bissell, 2010). The buzz of energy, the intensity of feeling, the waves of anxiety that may arise and then dissipate, can, as Conradson and Latham (2007) argue, creep through social encounters and movement through space. Out of these shifting arrangements arise particular dynamics of feeling, both in terms of affective fields and more personalized senses of emotion. Again, to quote Massumi: 'When I think of my body and ask what it does to earn that name, two things stand out. It moves. It feels. In fact, it does both at the same time. It moves as it feels, and it feels itself moving' (Massumi, 2002: 1).

Within the framing of *movement* and *feeling*, we do this within the spatial ecology of *place*. This is, as Conradson and Latham (2007: 238) outline, to incorporate the way that affective resonances can be found within 'the coming together of people, buildings, technologies and various forms of non-human life in particular geographic settings'. In other words, our placed surroundings become an integral part of how we, individually and collectively, *feel* and *move* through the world. Yet, at the same time, it is important to reflect critically on the disposition of the body, individual or collective, that Massumi relates to. The affective possibilities charged through our placed interactions with the world are not, as Conradson and Latham (2007: 238) go on to remind us, 'accessible to all', but are marked by cleavages that shape – and are shaped by – and axis of inclusion and exclusion.

Where, for Massumi (2002), affect gives primacy to the social and material parameters that enable us to make sense of our embodied and cognitive movements, perceptions and thoughts, this book foregrounds how intersections of *place* and *gender* are woven into our affective experiences, giving meaning to senses, such as inclusion and exclusion, safety and fear. In other words, it is not enough to say that affect is located within a framing of place (be it material, social, atmospheric or technological), but that both place and affect entwine to shape, and become shaped by, pre-reflexive interpretations of gender. As we move through space, we encounter and interact with surrounding frictions and responses that (re)produce affections, shaping affective understanding of gender. How place is constructed, maintained and encountered, both materially and symbolically, has profound implications for how masculinities and femininities, as well as cis-gender, trans, genderqueer and non-binary identities, are experienced as a system of uncoded and prelinguistic expressions, resonating between place, space and our gendered selves. Following Massey (2013) and other feminist geographers in the field, we see gender as culturally produced and as fluid, processual and multiplicitous. Woven into our placed surrounds, place and place making needs to be situated within a broader symbolic system of power; one which explicitly carries gendered affections of identity, exclusion and belonging.

Set against the centrality of place in experiences of gender disadvantage, our rationale is based on the power of affect to shape and be shaped by social relations, yet its underrepresentation in literatures on geographies of gender. As we move through and encounter place, we do not encounter neutral containers in which we write our social selves, but we engage and interact with material, symbolic and cultural orders of meaning, which gives resonance to what is *in* or *out* of place. Place, therefore, is the distillation of local and global processes that create, produce and shape varied orders of hierarchy and power, within which we engage with our surroundings, reproduce identities, emotions, attachments, security and belonging (Sack, 1999; Massey, 2013; McDowell, 2016). As a recent World Economic Forum report (2021) has evidenced, widening gender gaps persist across countries and regions, so that reducing gender inequality and understanding the complex interrelations between place and gender remain important local/global issues and major policy concerns. This has underpinned a growing literature on spatial dimensions of gendered advantage and disadvantage (for example, McDowell, 1999; Rose, 2003; Massey, 2013; McDowell, 2016). At the same time there has been an increasing interest in the role of affect in social relations of power and dominance within humanities and the social sciences, with emotions and affect becoming a significant theme in human geography since the early 2000s (Anderson and Smith, 2001; Davidson and Milligan, 2004; Davidson et al, 2006). However, the latter literature has largely been limited to the *emotional* dimensions of place (for example, the range of emotions evoked in different contexts), to the neglect of *affect* as a 'line of force' and 'disposition' that has a strong political and transformative potential (Thrift, 2004; Pile, 2009).

So, what of emotion and affect? For Clough (2008), emotion becomes the conscious narrative impacts of affect, while affect itself is about opening the body to the non-conscious, the never-to-be-conscious, pre-reflexive sensory experience of being in the world (Massumi, 2002). As we encounter the world, we may – and will – experience waves of emotion, from joy to anger to fear. Just as we might, when taking a wrong turn and encountering an unforeseen sense of threat, be jolted by feelings of anxiety and fear, we cannot think of ourselves outside of an intrinsic connection between movement and sensation, framed by affective encounters with the material, cultural and technological tapestry of the placed world around us (Conradson and Latham, 2007). Yet, it cannot be assumed that such processes are accessible, stable or equal for all. Rather, our dispositional characters will serve as frames through which we encounter the pre-conscious, non-linguistic and non-representational forces that shape – and are shaped by – placed ecologies of affect. Here, affect gives relational form to the different configurations of objects, be they material, technological, corporeal or social, that come together to shape movement, perception, thought and cultural practice as we move through the social and material world (Gregg and Seigworth, 2010; Massumi, 2010). How we feel about the world, be it consciously or unconsciously, tells us a great deal about how the world works (Åhäll, 2018). Sara Ahmed, in her influential *Cultural Politics of Emotion* (2014), discusses how emotions and affects are not simply in movement, but 'stick'. The politics of emotion, therefore, becomes a feminist framing of affect and an attempt articulate the whole-body, pre-reflexive experience of our gendered selves, the markers that 'stick' (as well as those that do not), and, like gender, plays a fundamental role in how the social world works.

In this way, an interest in affect necessarily involves a focus on bodies (human and non-human) and, with it, gendered pre-reflexive, whole-body codings of gender (Ahmed, 2014; Åhäll, 2018). Yet, as Conradson and Latham (2007) argue, like Brennan (2004), affect is transmitted through bodily – and gendered – encounters, but this experience cannot be separated from the placed social environment in which we are situated. Brennen (2004), similarly to Bøhling (2015), develops the notion of an 'affective atmosphere', a feeling that hits you when you enter into a room or a community, eliciting whole-body sensory emotions connected to identity and belonging. Each of these processes, bodies, gender and place, needs to be thought of as a whole affective relation, shaping and creating social experience and moods of encounter. Our different bodies will, after all, generate different affective responses with certain gendered, raced and sexed encounters entwining with the spatial and social frame around us to create meaning. As Åhäll (2018) maintains, it is because affective processes exist between and within bodies and space that affect has everything to do with gender. Gender, like space, after all, plays a fundamental role in how our social worlds work. Fundamentally, gender and place are experienced as a sensation, or an 'atmosphere' that, as Ahmed (2017: 27) puts it, exists as 'a gut feeling' with 'its own intelligence'. As Ahmed (2017: 27) concludes, 'you have to get closer to that feeling'.

It is here that this edited volume is situated, across the intersections between gender, place and affect and how this enables a deeper understanding of spatial and gender-based inequality – one which sees gender as a 'virtual' (Thrift, 2004) and corporeal affective process of becoming and possibility that is inseparable from context. Bringing affect theory to the analysis of gender and place, this book offers new critical insight into the dynamic process of interaction that flows through material and social architectures of place and our lived sense of self to produce affections of thought, feeling, emotion and moods. These include key affective-geographical formations of attachment, disruption and belonging, organizing themes within the collection, helping to construct gendered understanding of what it means to be in or out of place. The aims around which the volume is framed, are fourfold:

1. to introduce the importance of affect in understanding the everyday interactions between place settings and gendered experiences of self;
2. drawing on a relational ontology and with a focus on gender and affect, to offer new critical insight into the processual and interrelational dynamics of place;
3. to offer new critical insight into gender dynamics in relation to place by incorporating the significance of affective intensities, their embodiment and their circulation into understandings and experiences of gendered power;
4. to critically examine multiple and situated reverberances of affect and its impact in shaping meanings of masculinities and femininities across a range of critical contexts.

Working towards these aims and building critically applied knowledge that addresses a gap in gender studies and cultural geography, chapters in the book are divided into three interrelated sections.

Gender and attachment in places and spaces of work

Given that organizations and the work situation are likely major influences on how men and women experience affective relations of gender, place and space, our first section focuses on the employment context. As Darren T. Baker points out in Chapter 5, workplaces are 'deeply social and emotional spaces', where people make connections with others, experience success and pride, but also endure difficult moments, including the experience of inequalities and oppression. This section draws on key affective notions of attachment and detachment (for example, to/from norms, logics, objects and people) in terms of how people negotiate relationships within gendered workspaces and settings. In Chapter 1, 'The Affective, Gendered Processes of Place Making: Understanding the Home Conservatory as a Place of Artistic Work', Nick Rumens explores the home conservatory as a gendered and affective workspace in relation to his part-time occupation as a visual artist. As he points out, the historical association of the conservatory with the domestic sphere has implications for how it is understood as a gendered space. While domestic space has become attached to notions of private life, the family and, thus, of women and femininity, the organization of domestic space into temporal leisure and workspaces problematizes the notion of the space of the conservatory as 'feminine'. Deriving theoretical insights from Judith Butler's work on gender performativity, Nick explores in this fascinating personal account how conservatory space is performatively constituted as gendered workspace and how the interaction between the architecture of this ambiguous space and his sense of self as a visual artist produces affections of thought and emotion that constrain and enable him to 'feel his way' as a visual artist.

In Chapter 2, 'Postfeminism and Affect: Exploring the Affective Constitution of Postfeminist Subjectivities by Leaders in the City of London', Patricia Lewis brings together notions of postfeminism and place to explore the affective positivity and affective dissonance connected to the identity work of male leaders working in the City of London. As she persuasively argues, postfeminism has produced the cultural conditions which facilitate the reframing of leadership as 'contradictorily gendered', where culturally feminine relational behaviours are increasingly expected alongside attachment to masculine-marked practices. This compelling chapter draws on the concept of place and the specific cultural landscape of the City of London to develop a nuanced and context-specific analysis of postfeminism as a psychic and affective phenomenon and of its call to male leaders to invest in hybrid postfeminist subjectivities. In Chapter 3, 'Who's Afraid of Virginia Woolf: Affective Responses to Space, Objects and Atmosphere in a Writer's House Museum', Jessica Horne explores how historic house museums are affective spaces, offering potential for engagement with the past in the present (Gregory and Witcomb, 2008). This chapter draws on Sara Ahmed's (2010) understanding of 'happy objects' to explore volunteers' affective responses to space, objects and 'atmosphere' in the context of Monk's House, a historic house museum and National Trust property known for being the former home of the 20th-century novelist Virginia Woolf. The museum relies on the labour of volunteers, many of whom are volunteering in retirement. In this chapter, Jessica Horne argues that in being 'moved' by the house, garden and 'atmosphere', volunteers 'make' meaningful attachments to place for themselves and contribute to the remaking of gendered spaces at the museum (Ahmed, 2010).

In Chapter 4, 'Trading from Home: The Affective Relations of "Doing Finance" in the Domestic Setting of the Home', Corina Sheerin and Alex Simpson take contextual and spatial changes in work organization from COVID-19 to offer unique and timely insight into the challenges, possibilities and experiences of conducting a masculinist occupational endeavour in the largely feminized sphere of the home. Focusing on the City of London, they explore the affective relations between individuals and the material form that surrounds them through the experiential shift from the aggressive, fast-paced masculine sphere of the City to the domesticated domain. They reveal within this new blurred landscape tensions in gendered identity formation as hierarchies, social relationships, organizational norms and power dynamics are both upheld and reconfigured to elicit a heightened sense of belonging and to define what is considered, in affective terms, in and *out* of place.

Darren T. Baker, in Chapter 5, 'What Is the Potential of Psychoanalysis to Understand the Relationship between Space, Objects and Subject Formation?' draws on Winnicott's (2016) notions of 'transitional objects' and 'potential space' to offer a psychoanalytical reinterpretation of Tyler and Cohen's (2010) influential paper on women and space in organizations. Through his innovative analysis, Darren offers additional insights into the relationship between the internal self and how it is shaped reciprocally by objects and space. In particular, he highlights how identities, subjecthood and affective experiences are constructed through space as well as through attachment to objects, and how these play an important role in supporting women in the face of inequality and discrimination. As such, this insightful chapter draws on psychoanalysis to offer a deep understanding of how physical objects and space shape our affective life and workplace identities. In Chapter 6, 'Affecting a Desiring "Woman Worker": A Spatial Interpretive Ethnography of a Café in India', the final chapter in this section, Rajeshwari

Chennangodu and George Kandathil explore the processes of producing an urban café in India as a 'women-only' workspace that seeks women's empowerment. Drawing on Deleuze and Guattari's (1988) notion of desire as a connecting force and with a focus on affect as a material, embodied and interpersonal 'becoming', they skilfully delineate the processes that help to produce the café as a territorialized space. Through an interpretive ethnographic exploration, which includes participant observation of the kitchen space and the dining hall of the café, they reflexively analyse notions of 'women' and 'worker', exposing an affective hierarchy and attachment to the dominating neoliberal forces that prioritize logic of commodification over other logics and where women invest human bodies with an intense desire to become a 'better' woman worker.

Gender, disruption and unsettling spaces and places

Our second section is oriented around disruption and disturbance as well as around places as potential sites of insecurity and threat. As Sedgwick (2003) argues, not only do emotional connections threaten to disrupt dominant power relations, but the unpredictability inherent in affective relations also draws attention to the singular and the unexpected. Chapters in this section accordingly explore tensions and disturbances through gendered experiences and meanings associated with safety, home, exploitation and transgression. In Chapter 7, 'Taking Place in-as Soho: Understanding the "Here and There, Then and Now" of Gender and Affect Work', Melissa Tyler explores some of the ways in which working communities are composed of sociocultural, material and affective references that are spatially and temporally situated, enabling particular settings to 'take place'. Drawing on insights from feminist sociology, phenomenology and geography, she considers how Soho is performatively acted as a meaningful location. These meanings are shown to be masculine and hyper-heteronormative, and at the same time critically queer and multiplicitous, where the sociocultural, material and affective configurations of gender operate as sites both for exposing and for displacing the reified connotations of Soho as a place of gender exploitation. This powerful line of argument gives emphasis to Soho as a setting that brings together elements of its past, present and future (its 'then and now'), and of reference points within and beyond its physical and perceptual boundaries (its 'here and there'), into a constellation of affects, meanings and materialities that provide opportunities for gender to 'take place' in complex, contradictory and often critically reflexive ways.

In Chapter 8, 'Affective Practices and Liminal Space-making in Palestinian Refugee Camps', Alison Hirst and Christina Schwabenland examine bordering and gendering practices in the unique context of the Palestinian refugee camp, understood as 'edgelands', that is, spatially situated assemblages of unwanted, feared or stigmatized activities, things or people that are 'pushed to the outside'. Focusing on women's social enterprises, and drawing on new materialist notions of affect, they show how preparing and selling food makes it possible for women to transgress both domestic and camp boundaries and redefine their gendered identities. Through this rich and compelling case, they illustrate how assemblages combining human and non-human agencies (such as kitchens and digital images of food) enable women to construct liminal border zones in material and digital spaces and how, as spatial borders are remade and transgressed, gendered-identity norms concerning the nature and status of women's work are reconstituted. In Chapter 9, 'Placing Fear of Crime: Affect, Gender and Perceptions of Safety', Murray Lee offers a powerful insight into how men and women's affective engagement, encounters and resonances with places and spaces serve to non-consciously influence negative (or indeed positive) perceptions of safety. Here, Murray explores affect in gendered perceptions of safety, drawing on a range of empirical projects, and assesses how the environments we engage and interact with provide non-conscious cues and codes for individuals that serve to (re)produce hierarchies of safety and security. In doing so, the chapter sheds new light on gendered perceptions of fear, speaking to core issues and feelings of belonging and security as people move through and encounter different spaces within their everyday lives.

In Chapter 10, 'To Be a Homeless Woman in Russia: Coping Strategies and Meanings of "Home" on the Street', Evgeniia Kuziner explores the affective strategies of homeless women and how they use buildings and public spaces in Russia to construct a 'home' on the street. As she argues, there is an important gender dimension to the problem of homelessness, with women's homelessness remaining invisible, largely due to the particular stigma attached to the 'unaccommodated woman' as representative of a form of deviance. Homeless women therefore 'disappear' from the institutional spaces of homeless shelters and frequently rely on precarious arrangements to be 'housed'. Drawing on rich empirical data, the chapter explores the gender specifics and affective strategies of coping on the street and gendered differences afforded to meanings and constructions of 'home'.

Place, gender identity and belonging

Our final section draws on the socially constructed, the material and the visual/aesthetic aspects of place and how these are implicated in affective notions of belonging – our physical involvement with our environment and a key dynamic within the affective-geographical experience. Chapters in this section capture the critical role of collective and individual experiences of belonging as affective, localized formations, and how they translate into gendered, classed, raced manifestations of inclusion and exclusion, highly significant to subjectivity and our 'being in the world'. In Chapter 11, 'Affective Atmospheres of Finance: Gendered Impacts of Financialization within Sydney's Barangaroo Development', Alex Simpson and Paul McGuinness examine the impact of financialization on the material design and construction of the contemporary 'global city' and how the altered urban landscape shapes and genders social interaction in Sydney's Barangaroo development on the western waterfront of Sydney's central business district. The chapter explores the impact place making has on constructions of gender identity and the 'self' as well as on the affective thoughts, motions and rhythms of occupants moving through the placed environment. As they point out, at the centre of this, Barangaroo is an overwhelmingly 'neoliberal venture' that, as a gendered project, has profound implications for *who* and *how* individuals use space. The chapter and its case offer a unique opportunity to examine the strategic efforts made across state and private sectors to achieve the goal of a 'global city' as well as to analyse the critical implications this has for affective experiences of belonging and for gendered assumptions of urban design.

In Chapter 12, 'Liminality and Affect: Knowing and Belonging among Unscripted Bodies', Nyk Robertson explores relationships between affect and queer bodies and how these can bring about spaces of belonging and community. As they point out, queer bodies that do not follow prescribed scripts, often presented in a heteronormative and cisnormative way, must find possibilities in liminal spaces to be known and gain a sense of belonging. Through their in-depth analysis, they demonstrate how, while detachment comes at a cost, with associated feelings of loss, within liminal spaces, and within movement of spaces and identity, a freedom and hope of possibilities also exists. Queer bodies, therefore, have the potential to rewrite scripts and create possibility scripts that can generate space for other queer bodies, and queer connections within and between bodies and space. In other words, as they persuasively argue, while bodies are introduced to prescribed scripts though social structures, affect in queered spaces can expand possibilities and restructure social meanings. In Chapter 13, 'Unsettling Metronormativity: Locating Queer Youth in the Regions', Nicholas Hill, Katherine Johnson, Anna Hickey-Moody, Troy Innocent and Dan Harris take us through their fascinating 'methodological adventure' to report on their collaboration with LGBTIQA+ (lesbian, gay, bisexual, transgender, intersex, queer/questioning, asexual) youth living in a regional town near Melbourne, Australia. As they point out, limited acknowledgement of positive queer stories can transform feelings of place, connection and belonging and leave intact perceptions of rural and regional areas as unwelcoming and often hostile to LGBTIQA+ youth. Through their creative and innovative hybrid methodological approach, they highlight some hopeful stories of queer emplacement and belonging, thereby transcending normative understandings of queer lives as out of place within regional areas. As they argue, by examining the way LGBTIQA+ youth navigate regional and rural areas, new insights can be produced into the ways place, connection and belonging are enacted through everyday interactions and affective attachments to place.

In our final chapter, Chapter 14, 'Landscape, Gender and Belonging: Male Manual Workers in a UK Seaside Town', Ruth Simpson and Rachel Morgan explore the 'gendering' of landscape and how it is claimed through affective experiences of belonging. Drawing on a recent study of male manual workers in a 'struggling' UK seaside town, Hastings in East Sussex, the chapter highlights how the physical and cultural attributes of landscape help to generate attitudes of inclusion and exclusion through, in part, the claiming of landscape, as well as through ideologically charged and dominant ways of seeing and defining. The chapter explores gendered understandings of landscape and highlights how, in a recursive sense, it generates and reflects affective experiences of belonging, exclusive of 'outsiders', and which relate to how landscape is both inhabited and claimed.

★★★

Bringing affect theory to the analysis of gender and place, the volume offers new critical insight into the dynamic process of interaction that flows through material and social architectures of place and our lived sense of self to produce affections of thought, feeling, emotion and moods. These include key affective-geographical formations of attachment, disruption and belonging, organizing themes within the collection, helping to construct gendered understanding of what it means to be *in* or *out* of place. With no current text addressing this specific intersection, this timely edited collection builds on feminist geography (for example, McDowell, 1999; Rose, 2003; Bondi, 2004; Massey, 2013; McDowell, 2016) to offer an alternative vision of gender relations and place through a focus on embodied experiences and affective ties and how these have potential to cement as well as disrupt relations of domination and power. Furthermore, with few studies focusing specifically on place (Hopkin, 2018), the application of an intersectionally attentive framework introduces the role of locality and affect into how axes of privilege and oppression (for example, based on gender, class, sexuality, race) are recursively shaped. In short, as the fields of geographies of gender and literature of inequality and affect become more diverse, this text leads the way by bringing together key strands of thought to shape how we understand gender and 'our place in the world'. By engaging such themes, this edited volume advances critical debates surrounding the gendering of place, symbolic manifestations of inclusion and exclusion as well as, in affect theory, bringing a new approach to the core notion of spatiality as a product of gendered relations. After all, it is important to remember, as we move through and encounter place, we do not encounter neutral containers in which we write our social selves, but we engage and interact with material, symbolic and cultural orders of meaning, which gives resonance to what is *in* or *out* of place.

References

Åhäll, L. (2018) Affect as Methodology: Feminism and the Politics of Emotion. *International Political Sociology*, 12(1): 36–52.

Ahmed, S. (2010) Happy Objects. In: M. Gregg and G. Seigworth (eds) *The Affect Theory Reader*. Durham, NC: Duke University Press, pp 29–51.

Ahmed, S. (2014) *The Cultural Politics of Emotion*. Edinburgh: Edinburgh University Press.

Ahmed, S. (2017) *Living a Feminist Life*. Durham, NC: Duke University Press.

Amin, A. and Thrift, N. (2002) *Cities: Re-imagining Urban Theory*. Cambridge: Polity.

Anderson, K. and Smith, S. (2001) Editorial: Emotional Geographies. *Transactions of the Institute of British Geographers*, 26(1): 7–10.

Bissell, D. (2010) Passenger Mobilities: Affective Atmospheres and the Sociality of Public Transport. *Environment and Planning D*, 28(2): 270–289.

Bøhling, F. (2015) The Field Note Assemblage: Researching the Bodily-Affective Dimensions of Drinking and Dancing Ethnographically. In: B. T. Knudsen and C. Stage (eds) *Affective Methodologies: Developing Cultural Research Strategies for the Study of Affect*. Basingstoke: Palgrave Macmillan, pp 161–181.

Bondi, L. (2004) Locating Identity Politics. In: M. Keith and S. Pile (eds) *Place and the Politics of Identity*. London: Routledge, pp 89–106.

Brennan, T. (2004) *The Transmission of Affect*. Ithaca: Cornell University Press.

Clough, P. T. (2008) The Affective Turn: Political Economy, Biomedia and Bodies. *Theory, Culture and Society*, 25(1): 1–22.

Conradson, D. and Latham, A. (2007) The Affective Possibilities of London: Antipodean Transnationals and the Overseas Experience. *Mobilities*, 2(2): 231–254.

Davidson, J. and Milligan, C. (2004) Embodying Emotion Sensing Space: Introducing Emotional Geographies. *Social and Cultural Geography*, 5(4): 523–532.

Davidson, J., Bondi, L. and Smith, M. (2006) *Emotional Geographies*. London: Routledge.

Deleuze, G. and Guattari, F. (1988) *A Thousand Plateaus: Capitalism and Schizophrenia*. London: Bloomsbury Publishing.

Gregg, M. and Seigworth, G. J. (2010) An Inventory of Shimmers. In: M. Gregg and G. J. Seigworth (eds) *The Affect Theory Reader*. Durham, NC: Duke University Press, pp 1–25.

Gregory, K. and Witcomb, A. (2007) Beyond Nostalgia: The Role of Affect in Generating Historical Understanding at Heritage Sites. In: S. Knell, S. MacLoed and S. Watson (eds) *Museum Revolutions: How Museums Change and Are Changed*. Canada: Routledge, pp 263–275.

Halford, S. and Leonard, P. (2006) Place, Space and Time: Contextualizing Workplace Subjectivities. *Organization Studies*, 27(5): 657–676.

Hopkin, P. (2018) Feminist Geographies and Intersectionality. *Gender Place and Culture*, 25(4): 585–590.

Massey, D. B. (2005) *For Space*. London: SAGE.

Massey, D. B. (2013) *Space, Place and Gender*. Bristol: Policy Press.

Massumi, B. (2002) *Parables for the Virtual: Movement, Affect, Sensation*. Durham, NC: Duke University Press.

Massumi, B. (2010) The Future Birth of the Affective Fact: The Political Ontology of Threat. In: M. Gregg and G. J. Seigworth (eds) *The Affect Theory Reader*. Durham: Durham, NC: Duke University Press, pp 52–70.

Matthews, D. and Wall, I. (2023) Legal Aesthesis: Affect, Space and Encounter. *Law, Culture and the Humanities*, 19(2): 186–190.

McDowell, L. (1999) *Gender, Identity and Place: Understanding Feminist Geographies*. Bristol: Policy Press.

McDowell, L. (2016) *Migrant Women's Voices: Talking About Life and Work in the UK since 1945*. London: Bloomsbury.

Pile, S. (2009) Emotions and Affect in Recent Human Geography. *Transactions*, 35: 5–20.

Rose, G. (2003) *Feminism and Geography: The Limits of Geographical Knowledge*. Bristol: Policy Press.

Sack, R. (1999) A Sketch of a Geographic Theory of Morality. *Annals of the Associations of American Geographers*, 89(1): 26–44.

Sedgwick, E. K. (2003) *Touching Feeling: Affect, Pedagogy, Performativity*. Durham, NC: Duke University Press.

Thrift, N. (2004) Intensities of Feeling: Towards a Apatial Geography of Affect. *Geografisko Annaler Series B*, 86: 57–78.

Tyler, M. and Cohen, L. (2010) Spaces that Matter: Gender Performativity and Organizational Space. *Organization Studies*, 31(2): 175–198.

Winnicott, D. W. (2016) Transitional Objects and Transitional Phenomena. In: L. Caldwell and H. Taylor Robinson (eds) *The Collected Works of D. W. Winnicott: Volume 3, 1946–1951*. Oxford: Oxford University Press.

World Economic Forum (2021) *Global Gender Gap Report 2021* Available at: https://www3.weforum.org/docs/WEF_GGGR_2021.pdf

Young, A. (2023) Arrested Mobilities: Affective Encounters and Crime Scenes in the City. *Law, Culture and the Humanities*, 19(2): 210–226.

PART I

Gender and Attachment in Places and Spaces of Work

1

The Affective, Gendered Processes of Place Making: Understanding the Home Conservatory as a Place of Artistic Work

Nick Rumens

Introduction

Stereotypes of visual artists working in industrial lofts or airy studios, bestrewed with tubes of paint, brushes, easels and furniture, in up-and-coming urban areas within cities and towns, persist, despite being critiqued as overstated and out of touch with the lived realities of where artists work (Florida, 2002; Markusen, 2013; Hall, 2022). As Markusen (2013: 524) points out, such stereotypes 'obscure the presence and roles of artists in very small communities … in smaller and midsized towns'. Indeed, numerous artists are located away from the heart of metropolises, in suburbia, provincial towns and (semi-)rural locations, making art in places and spaces not originally designed and organized for carrying out paid artistic work. One of these places is particularly complex, that being the home, because of its associations with unpaid domestic work and its attraction as a place that, traditionally, is separate from the realm of paid employment (Halford, 2006; Taylor, 2015). While this image has been shattered by research about the home as a production site of economic activity (Felstead and Jewson, 2000; Christensen, 2019), the home retains strong connotations as a place that is free from the conventions and binds of the workplace outside the home, where personal values and preferences can be prioritized (Taylor, 2015). This can present several challenges for artists engaged in paid artistic work. Artists working in spaces of the home may have to renegotiate the associations they harbour as gendered spaces that pertain to domesticity, leisure and family. As such, paid artistic work may be difficult to perform in home spaces where domesticity and family prevail, just as some home-based work can encroach into or 'invade' spaces that are specified for family and leisure activities (Taylor and Littleton, 2016; Wapshott and Mallett, 2012). Reconfiguring the home as a place for paid work, some artists, including myself, have (re)constructed spatial boundaries between work and non-work, partitioning domestic spaces into something that approximates a studio space purposed for art making, with varying degrees of success. Objects may also be important in that endeavour, such as tubes of paint, easels, brushes and canvas frames, as these artefacts help to replicate the image of what a professional artist's studio ought to look like. Considering this, I focus this chapter on exploring the gendered dynamics and affective engagements in reconfiguring non-workspace within the home so it may be experienced as a place of work.

Pertinent to this chapter is the scholarship of human geography, which has amply demonstrated how places are more than physical locations but are constituted as such through processes of place making that are linked to human experience (Cresswell, 2015). Ontologically, place is understood in this chapter as a sense of place that refers to subjective and affective attachments that people develop to a place, which are enacted over time through experience and the past, current and projective meanings that coalesce around a place. There is a pronounced emphasis in human geography on how place making is bound up with the indeterminacy of affect, as a 'state of being' (Hemmings, 2005: 551) and relational force that transits through spaces and bodies (Duff, 2010; Gregg and Seigworth, 2010). Deriving conceptual insights from theories of affect, geography scholars have highlighted affect as a device that illuminates the states of being, encounters and engagements involved in place making, which can enhance people's sense of belonging toward a specific place (Duff, 2010). Additionally, processes of place making are intertwined with gender, where gender produces shaping effects on how people understand and experience place in terms of gender relations, identities and subjectivities (Massey, 2013). The gendering effects of place are significant in different ways, not the least of these being how they can shape pathways to employment, revealing how places are understood as sites of work (Simpson et al, 2021, 2022).

Specifically, I write about place making as a man who is a paid academic, but also a self-employed visual artist, whose artwork is produced in a corner of a home conservatory. Prior to this, I worked as a full-time academic, but my late mother's ailing health and associated caring responsibilities were one of several reasons for continuing academic employment on a part-time basis. While this represented a significant loss of earnings, the available time outside of paid academic employment and unpaid caring work, combined with support from my partner, meant I could pursue a Master's degree in

Fine Art Painting. Initially, it was not my intention to pursue artistic work on a paid basis, but through a number of serendipitous events my paintings began to sell, and my profile as a visual artist started to escalate. That being said, my experience of paid artistic work chimes in with the experiences of other visual artists documented in research (Alacovska and Bille, 2021), which is that it is intermittent, uncertain and poorly paid, and requires me to juggle multiple work roles. Currently, the portion of time I allocate to paid artistic work fluctuates in line with the shifting and at times overbearing demands of paid academic work, the latter being a vital source of fixed income. It is important for me to acknowledge that my part-time status as a self-employed visual artist alters the relationship I have with my studio space, not least because it limits the time I can commit to it as affective workspace in the making. For this reason, and financial ones also, I do not have a dedicated studio space that is separate to the home or a room in the home that may be used only for this work. Indeed, my 'studio' is far removed from popular images of painters occupying loft spaces, warehouses, or rented spaces in community art-making centres (Hall, 2022). Indeed, it barely deserves the label of 'studio' in the traditional, masculine sense; rather, I understand and experience it as a studio space that is coded as feminine, which supports my painting practice. This space is a distinct, relatively small corner of a home conservatory, surrounded by portable storage trolleys, and contains two easels, copious bottles of painting medium, paint pots, brushes and house plants, with a dust sheet on the floor (Figure 1.1). Importantly, it is not a passive or a singular, static space, but comprises multiple spaces that are relational and intermeshed with spaces for socializing, relaxing and drying laundry. Viewed in this way, my studio space has competing meanings attached to it that relate to the gendered dynamics of home life as well as the new ones that connect to a place of paid artistic work.

To begin my explorations, I outline how affect is deployed in this chapter, before briefly discussing the organizational literature on affect, place and space. Next, I delve into the cultural history of the home conservatory to show how place making is shaped by the layered meanings spaces can accrue over time. While my home conservatory is a relatively new structure, its place and meaning in the history of the conservatory are carried over in my affective engagements with its spatial dynamics as a place of work. Extending my discussion, I locate the enduring image of the artist's studio as masculine in the cultural history of the artist's studio, before articulating the positive affective states within my studio space that are coded in femininity. In another way, the home conservatory retains associations as a gendered space for domestic and leisure activities, and these permeate processes of place making that can undermine the affective states of being that enable me to undertake paid artistic work. Rounding off, the chapter finishes with some concluding remarks.

Figure 1.1: Studio space in a home conservatory

Source: Nick Rumens

Affect, gender and places and spaces of work

To begin unpacking the interrelationship between places of work, affect and gender, it is useful to consider first the concept of affect. Primarily, I connect with geography and cultural studies scholars who converge on the view that there is no general consensus on what affect means. Indeed, one persuasive perspective is that a general theory of affect would foreclose its capacity to shift across disciplines and diminish the diversity in how it has been and continues to be used (Massumi, 2002; Thrift, 2004, 2007; Seigworth and Gregg, 2010). Yet over time theories of affect have accumulated a bibliographic shape (Seigworth and Gregg, 2010), which permit us to say something about what affect *does* rather than definitively deciding what it is and is not. Seigworth and Gregg (2010: 1–2, emphasis in original) hold that affect is a '*force or forces of encounter*' that exceeds the limits of emotion in how they drive individuals 'toward movement, toward thought and extension'. Similarly, Hemmings (2005: 551) submits that affect refers to 'states of being, rather than to their manifestation or interpretation as emotions'. Likewise, I conceptualize affect as a relational force that occurs in and disseminates through places and spaces. In this way, affect is not reducible to a specific place, space or body, but is experienced as a state of being or as an intensity that shapes the modes of existence bodies may take within and across specific places. It is in that sense that the body has the capacity to affect as well as be affected. What may be gained by adopting this perspective is recognition and appreciation of how affect can shape actions and people's dispositions and orientations to the world within specific spaces and places (Duff, 2010). Here, affect may be understood as a state of being and relational force that plays a constitutive role in place making, in that to experience place is to be '*affected by place*' (Duff, 2010: 881, emphasis in original). Crucial also is how we can understand affect to attach itself to objects, people, spaces, places, ideas and so on (Sedgwick, 2003; Hemmings, 2005), which forces scholars to pay close attention to affects in everyday life. Notably, affects and affective states are qualitatively ambiguous and unpredictable, difficult to pinpoint or pin down, such that individuals may not be conscious of their origins or able to describe their character (Thrift, 2004). Understood in this way, exploring affect can cast light on its slipperiness and diffusion through spaces and places, as well as the numerous, often unanticipated attachments it forms in the processes of place making.

While scholarly interest in affect in the humanities and social sciences acquired significant momentum from the 1990s, the same cannot be said about organization studies. In similar veins within which affect has been conceptualized by theorists of affect cited earlier, so it has among some organizational researchers over the last few decades. Noting the dearth of research on affect in organization studies, Fotaki et al (2017: 4) comment similarly on affect's elusory quality: 'it [affect] is about an aspect of bodily experience that eludes interpretation by language, escaping its logic and refusing to conform to its expectations'. Embracing its indeterminate quality, organizational scholars have deployed affect to analyse management learning as an affective process (Beyes and Steyaert, 2021), study the dynamics between affective atmospheres and organizing (Michels and Steyaert, 2017), affective suffering and work (Dashtipour and Vidaillet, 2017), and affect and gender (Pullen et al, 2017). These valuable contributions have drawn affect into organizational analyses, but this strand of literature has yet to explore fully the relationship between affect, work and place making.

Place making has aroused attention among some researchers in the context of work environments (Lawrence and Dover, 2015; Dacin et al, 2018; David et al, 2020; Cartell et al, 2022) and employment opportunities in places such as UK seaside towns (Simpson et al, 2021, 2022). This research underlines the importance of place making as a process through which people come to experience places in terms of work and employment. Cartel et al (2022) argue that place making is a profoundly subjective experience that is pivotal to the maintenance of place over time, and historical meanings that orbit around a place of work in the past may shape meanings attributed to place in the present and the future. Other studies have teased out how place and gender are enmeshed in ways that contour, for example, how working-class men perceive and experience places of work and the employment pathways available to them (Simpson et al, 2021, 2022). Research in this domain has yet to examine places of work located within the home, to explore the affective aspects of place making. The analysis of spaces of work forms part of a wider segment of literature on organizational space, which has been conceptualized from a variety of theoretical angles (Conradson, 2003; Tyler and Cohen, 2010; Dale et al, 2018). One notable area of research has opened critical encounters with the spatial implications of home-based work (Halford, 2006; Waismel-Manor et al, 2021). Wapshott and Mallett (2012) examine the collapsible spatial demarcations in the home/work environment. People may strictly delineate home spaces for work by, for example, designating a room as an 'office space' and equipping it with objects associated with office work (for example, computers, telephones, filing cabinets). Reconfiguring domestic space in this manner does not mean the work/non-work boundary remains intact. Inhabitants of home spaces may be noisy and engage in leisure pursuits in spaces earmarked for work, interrupt the home worker with matters pertaining to domestic life, all of which demonstrate the permeability of spatial borders that divide work and non-work. Meanings attached to new spaces of work in the home may not be shared by all home occupants, highlighting how the meanings associated with domestic spaces (for example, in terms of family and leisure) can intrude and are not always easy to contain or discard.

The gendered dynamics of such spaces, overlooked in Wapshott and Mallett (2012), have been brought to the fore by other researchers (Halford, 2006; Waismel-Manor et al, 2021; Islam, 2022). Analyses have shown how spaces of the home for paid work are embedded in and constituted through gendered power relations that encode work and domestic spaces as masculine and feminine, respectively. The constitution of work/home spaces through gender norms that sustain a gender binary (masculine/feminine, man/woman) can be undercut by competing needs and demands of work and domestic life, such as when women feel compelled to undertake domestic work during time allocated for paid work, exacerbating extant gender inequalities (Islam, 2022). Demonstrating this also, Waismel-Manor et al's (2021) study of men and women spouses' working patterns in the home showed an unequal negotiation of space and time in the home, where men had more uninterrupted time and allocated space than women, whose workspace and time were more fragmented.

In summary, organization studies literature provides an array of valuable but disparate insights into the dynamics between affect, place making, gender and space. Accordingly, I weave together scholarly threads from human geography and the research areas within organizational studies cited earlier. In so doing, my explorations in the following add another dimension to extant debates on place making, gender and affect by considering paid artistic work within the spaces of the home.

The gendered cultural history of the conservatory

Through my ongoing efforts to experience a corner of my home conservatory as a place of work, I have at times been cognisant about how processes of place making are saturated with past, present and projected meanings that relate to the non-workspaces of the home. Contemporary home conservatories have their origins in early and relatively simple fabricated structures that sprouted up in the 16th century in the UK and across Europe to protect plants from harsh weather conditions. These constructions served similar purposes to the greenhouses and glasshouses with which we are currently familiar, such as those used for indoor gardening (Ruppel, 2020). The early conservatories – sometimes called orangeries – were very expensive to construct, not least due to the high costs of materials such as glass, window taxes and the associated expense of heating these structures, which meant they could be financed and built only by the wealthy. These typically standalone conservatories housed frost-tender tropical plants and citrus fruits that had been collected from overseas expeditions to the Mediterranean. Other conservatories housed the botanic spoils of colonialism such as rubber and coffee-bean plants. Some of the early conservatories were constituted as places for the dissemination of scientific knowledge about plants. Yet, the sheer novelty of some of these plants, as symbols of the 'tropical', such as the banana plant (Arnold, 2006), attracted excitement and widespread attention. Conservatory owners invited their peers, dignitaries and prominent social figures to peruse their exotic plant collections. It is easy to imagine how visitors may have been affected by these plants, and generated affective encounters and states of being from viewing some of the first tropical plants imported to western shores from the, not unproblematically termed, 'new world'. These affective engagements would have been shaped by gender. As scholars have noted (Schiebinger, 2007), plants in botanic conservatories helped to reproduce imperial discourses of European colonial masculinity as a 'positive' and 'progressive' force. In this frame, affective processes of place making are interwoven with colonial masculinity and men's practices, which orients our consideration of affect and affective states toward how they are implicated in the gendering of place.

With its roots firmly embedded in botany and colonialism, the conservatory became a signifier of an aspirational and desirable place to occupy. By the mid-19th century, referred to as the 'golden age of conservatories', advancements in heating technology as well as metal and glass production were such that conservatories became more available to people. Conservatories were attached to houses, and the resulting spatial arrangements, as being neither fully inside nor outside, made conservatories ambiguous spaces that opened up alternative modes of use. The conservatory continued to satisfy the pursuit of horticultural interests, which had become almost feverish in the 19th century, enabling conservatory owners to indulge in collecting diverse species of ferns and flowering plants (Thompson and Borozdin-Bidnell, 2019). Significantly, the conservatory developed as a desirable place for entertaining, dining as well as familial and romantic intimacy.

At this juncture I turn to examples of artwork that show the Victorian conservatory as a reconfigured space for bourgeois women and their interests and pastimes. One motive behind this is that affect attaches to artworks, such as when a painting opens a portal for viewers to experience the world differently. Works of art can create places – such as the interiors of home conservatories – that invite us into the realm of affect (Schilo, 2016). We may experience this as an 'excess' in how we are moved by what we see, experiencing the artwork 'beyond its material form' and the visual language it deploys through paint (Schilo, 2016: 122). In that respect, glimpses into the gendered dynamics and affects within and shaping conservatory spaces are palpable in the paintings produced at and around this time. Eduard Gaertner's *The Family of Mr Westfal in the Conservatory* (1836), Harry E. J. Browne's *Tea in the Conservatory* (c. 1870), Edouard Manet's *In the Conservatory* (1879) and Henri Matisse's *The Conservatory* (also known as *Two Female Figures and a Dog*, 1937–38) render the conservatory as a space in which women are seen to socialize, take afternoon tea, play with their children and relax. I experienced the excess of affect the first time I saw Gaertner's painting *The Family of Mr Westfal in the Conservatory* (1836), which depicts Mrs Westfal and her three daughters in a spacious conservatory. At the centre of the painting is a dining table with a tea spread, at which Mrs Westfal is sat, touching one of her daughters. The other two are positioned in front of her, playing, while behind Mrs Westfal is a vast conservatory wall adorned with numerous shelves of colourful exotic plants. These plants and the creeping vine ascending to the apex of the conservatory roof almost overshadow Mrs Westfal and her daughters, reasserting the meanings associated with conservatory spaces for the display of exotic and ornamental plants. Gaertner's painting constructs the conservatory space at that time as a charming, refined and intimate extension of domestic space that is detached from the world of work occupied by men. While paintings of conservatory spaces from the late 18th to mid-19th century tended to figure women in reclining positions or engaged in family activities or pastimes such as reading and painting, some paintings articulated the more ambiguous aspects of conservatory space. As Burton (2015) notes of the conservatory paintings of Joseph Tissot (1836–1902), the porous and liminal spatial dimensions of the conservatory as an in-between inside-outside space help to convey the risks and ambiguity bourgeois white women were likely to encounter in rivalling for the attention of a potential suitor (for example, *In the Conservatory [The Rivals]*, c. 1875–88), or flirting with an unseen admirer as in *The Fan* (c. 1875). In these and other Tissot paintings, women can be read as being highly skilled in capturing the attention of men, enticing them into the lush and shadowy corners of the conservatory.

Returning to my home conservatory, I am struck by how far removed it at first appears from its earlier Victorian incarnations and cultural representations; but, on reflection, enacting a sense of place that relates the conservatory to paid work, it inherits a weighty history of meanings that have over time been reorganized through affective processes of place making. In other words, the modern home conservatory evokes memories of the past that bear down on my sense of place making in the present (Cartel et al, 2022). I have partitioned off a corner of the conservatory for the purpose of paid artistic work, informed my partner this is a space for work, and incorporated objects (for example, painting tools) that signify it as such. However, I am aware that my conversion of domestic space to meet the needs of paid work has an affective afterlife. The conservatory is also constituted as a space for indoor gardening and leisure, as well as drying laundry. There are 25 or more pot plants that line the conservatory windowsill and the tops of two storage units. With a nod to the past when conservatories housed unusual and ornamental plants, the plants in my conservatory have been selected for their ornamental aesthetic as well as for their strange forms and sinister behaviour, exemplified by a healthy collection of carnivorous plants (pitcher plants, sundews and butterworts), specialist orchids and rare succulents and cacti. These plants are important visual references for my artmaking, which aims to develop pictorial languages that enable the viewer to experience representations of plant life beyond the forms they may already be familiar with. It has occurred to me that these plants are heavily implicated in the affective processes of place making.

Affect attaches to objects such as plants, whereby I experience a state of being that moves me toward a vision of the plant kingdom that is out of kilter with and beyond the materiality of my own plant collection. Of course, there are other possible affective engagements with these plants. As one friend once exclaimed, after shuddering at the sight of one creeping pitcher plant scaling the conservatory window, 'I don't know why, but they unhinge me'. Plants can be experienced as monstrous and unattractive, just as they can affect us to the contrary. The affective engagements I experience with my conservatory plants exert a constitutive effect in how the conservatory is understood as a place of paid artistic work. This is evident in how I interact with the plants as 'visual references', essential to the development of my painting practice that is a mode of plant-themed art, rather than tending to them solely as passive, aesthetically beautiful objects. In this way, my affective interactions with the plants position them as active participants in my enactment of place that help me to establish attachments to the conservatory as a workspace. But the affects attached to plants can produce alternative meanings and affective states. Festooning spaces in the conservatory beyond my partitioned space of work, the plants evoke historical memories of place in how they gender the spatial dynamics of the conservatory as a feminine place for pursuing the pastime of indoor gardening. This is a pastime I undertake within the conservatory. In this way, enactments of place making that may attempt to (re)imagine the conservatory as a studio space for artistic paid work that, traditionally, has cultural associations with masculinity (Schilo, 2016), are interrupted. This is elaborated in the next section.

The gendered cultural history of the artist's studio

As discussed earlier, the spatial and affective dynamics of the home conservatory are such that it houses competing meanings that relate to the gendering of place. Enlarging on this, some observations pertaining to the cultural history of the artist's studio are informative. The enduring image of the artist's studio is characterized as one that has long been associated with men, masculinity and men's work practices (Taylor, 2015). Hall (2022: 28–29) notes that the origins of the artist's studio can be traced to ancient Greece, which established the beginning of the idea of the artist's 'workspace as art gallery … seminar room and visitor attraction'. These early accounts of the artist's studio as a preserve of men set the tone and basis out of which a potent masculine image of the artist's studio crystallized. This can be traced to the 20th century and is discernible in the present. America painter Jackson Pollock (1912–56) and his wife, Lee Krasner (also a prominent painter), purchased a farmhouse and converted the barn into a studio space wherein Pollock could create his famous splatter paintings. Andy Warhol (1928–87) rented an unoccupied loft space before moving to another located on the fifth floor at 231 East 47th Street in Midtown Manhattan (1964–68), which became his first famous 'factory' for creative work. Anselm Kiefer (born 1945) took over an abandoned silk factory near Paris in 1992 and, over the decades, constructed a vast 200-acre compound comprising multiple studio spaces in engineered barns, towers, tunnels, greenhouses and lead-lined rooms. Kiefer stretches the idea of the artist's studio to an extreme, not least as an exercise of hyper masculinity that is writ large on the landscape. Artist studios have been represented and written about and critiqued over the years as places associated with the reproduction of masculinity that has excluded women artists (Bergstein, 1995; Thomas, 2015; Ringelberg, 2017), as well as spaces in which men artists work in isolation, eat, sleep and interact with potential buyers (Hall, 2022). Despite the diversity in contemporary working arrangements for paid artists (Florida, 2002; Markusen, 2013; Hall, 2022), the association between the artist's studio and masculinity endures, and is mobilized as a resource for identifying as a professional artist. As in Bain's (2004, 2005) research, Canadian women working as professional visual artists reported that an artist's studio was essential for being taken seriously in that regard, because of its attachments to masculinity and work.

The gendering of the artist's studio as masculine is interwoven with the affective processes of place making. For the most part, I find it challenging to perceive and sustain the experience of studio space as a masculine place of work. As I have discussed earlier, the enactment of place can occur by reproducing traditions that link conservatory spaces to their past (Cartel et al, 2022), which reassert it as a place coded as feminine. One option might be to repudiate this femininity, in order to try to sever spaces of the conservatory from meanings of domesticity to project new meanings of paid work, which can be linked to masculine images of the 'virile' professional man as paid artist. Instead, some of my enactments of place are structured by a template of plant–human engagement that is unafraid to acknowledge its cultural history as a mode of visual art that in the historical past has been allocated a lowly status, coloured by outmoded stereotypes of women as flower painters (Swinth, 2001). I feel most rooted in the spaces of the conservatory when I'm involved in an art of discerning available affective cues to foster my experience of the conservatory as a place to support my artistic work into the plant realm. How affect variously attaches to plants, as I have already mentioned, can move me to approach painting as a form of research into the plant world upon which we rely in order to thrive. The gendering effects of place, through which the home conservatory is experienced as spaces coded in femininity, based on leisure and pastimes such as indoor gardening, nourishes rather than impoverishes my sense of self as a plant artist. Expressed differently, these affective attachments galvanize my potential for action, to develop practices of painting.

However, affective engagements with objects in the conservatory have gendering effects that can stifle the potential for action. For example, as depicted in Figure 1.1, a laundry-drying rack can be seen from the right-hand side. This is a common occurrence during those months when washing cannot be dried outside. The drying rack gets moved around the conservatory and sometimes, such as when other housework chores are being undertaken (for example, vacuuming the conservatory floor), it can be unintentionally repositioned, so it impinges upon the studio space.

Notably, affect attaches to the drying rack in how it generates a state of being in me that is unconducive to painting. When I have planned to work on a painting and, upon entering the conservatory, I encounter the drying rack near or intruding into the studio space, I have automatically turned around and left the conservatory in frustration, questioning my sense of the conservatory as a place of work. In these instances, when meanings of domesticity circulate around objects such as a drying rack, my experience of the conservatory is a place in which domesticity presides. Crucially, such affective engagements can foreclose the possibilities for action, emphasizing a present orientation toward artistic work whereby it is abandoned. As such, objects can interrupt the experience of the conservatory as a positive place coded in femininity for creating plant-themed artwork. The presence of the drying rack re-establishes meanings that link spaces of the conservatory with domesticity – again coded as feminine – but recoded negatively as a place in which the everyday actions of housework undermine its sense of place that is related to paid artistic work.

Conclusion

In this chapter I have explored the affective, gendered processes of place making as they relate to the reconfiguration of home spaces into spaces that constitute a place of paid artistic work. In so doing, I hoped to have shown some of the affective states and encounters that affect the experience and enactment of place that are entangled within gendered meanings of work and non-work. Place making can be coloured by past, present and projected meanings of place that are layered and accrue over time, some of which persist and cannot be easily contained by partitioning space into work and non-work zones. As the foregoing discussion shows, place making can be experienced as a series of ongoing affective ebbs and flows that are sensitive to the dynamics of gender, whereby the historical and cultural associations of the home conservatory as feminine can support and detract from it as a place of paid artistic work that is plant centred. This represents an important area of organizational analysis that deserves future research, building on extant studies on how and where artists work, and how they are affected by places of work that are in the home. Affect could be mobilized by organizational researchers to construct inroads into place making to highlight how affect diffuses through places of work, shaping states of being that may be experienced, both positive and negative, for undertaking paid artistic work.

References

Alacovska, A. and Bille, T. (2021) A Heterodox Re-reading of Creative Work: the Diverse Economies of Danish Visual Artists. *Work, Employment and Society*, 35(6): 1053–1072.

Arnold, D. (2006) *The Tropics and the Traveling Gaze: India, Landscape, and Science, 1800–1856*. Seattle: University of Washington Press.

Bain, A. L. (2004) Female Artistic Identity in Place: The Studio. *Social and Cultural Geography*, 5(2): 171–193.

Bain, A. (2005) Constructing an Artistic Identity. *Work, Employment and Society*, 19(1): 25–46.

Bergstein, M. (1995) The Artist in His Studio: Photography, Art, and the Masculine Mystique. *Oxford Art Journal*, 18(2): 45–58.

Beyes, T. and Steyaert, C. (2021) Unsettling Bodies of Knowledge: Walking as a Pedagogy of Affect. *Management Learning*, 52(2): 224–242.

Burton, S. (2015) Champagne in the Shrubbery: Sex, Science, and Space in James Tissot's London Conservatory. *Victorian Studies*, 57(3): 476–489.

Cartel, M., Kibler, E. and Dacin, M. T. (2022) Unpacking 'Sense of Place' and 'Place-making' in Organization Studies: A Toolkit for Place-sensitive Research. *Journal of Applied Behavioral Science*, 58(2): 350–363.

Christensen, K. (2019) *The New Era of Home-based Work: Directions and Policies*. London: Routledge.

Conradson, D. (2003) Doing Organisational Space: Practices of Voluntary Welfare in the City. *Environment and Planning A*, 35(11): 1975–1992.

Cresswell, T. (2015) *Place: An introduction* (Second edition). J. Wiley and Sons.

Dashtipour, P. and Vidaillet, B. (2017) Work as Affective Experience: the Contribution of Christophe Dejours' 'Psychodynamics of Work'. *Organization*, 24(1): 18–35.

Dale, K., Kingma, S. F. and Wasserman, V. (eds) (2018) *Organisational Space and Beyond: the Significance of Henri Lefebvre for Organisation Studies*. London: Routledge.

David, R., Jones, C. and Croidieu, G. (2020) Special Issue of Strategic Organization: Categories and Place: Materiality, Identities, and Movements. *Strategic Organization*, 18(1): 245–248.

Duff, C. (2010) On the Role of Affect and Practice in the Production of Place. *Environment and planning D: Society and Space*, 28(5): 881–895.

Felstead, A. and Jewson, N. (2000) *In Work, at Home: Towards an Understanding of Homeworking*. London: Routledge.

Florida, R. (2002) Bohemia and Economic Geography. *Journal of Economic Geography*, 2(1): 55–71.

Fotaki, M., Kenny, K. and Vachhani, S. J. (2017) Thinking Critically about Affect in Organization Studies, Why it Matters. *Organization*, 14(1): 3–17.

Gregg, M. and Seigworth, G. J. (eds) (2010) *The Affect Theory Reader*. Durham, NC: Duke University Press.

Halford, S. (2006) Collapsing the Boundaries? Fatherhood, Organization and Home Working. *Gender, Work and Organization*, 13(4): 383–402.

Hall, J. (2022) *The Artist's Studio: A Cultural History*. London: Thames and Hudson.

Hemmings, C. (2005) Invoking Affect: Cultural Theory and the Ontological Turn. *Cultural Studies*, 19(5): 548–567.

Islam, A. (2022) Work-from/at/for-home: COVID-19 and the Future of Work – a Critical Review. *Geoforum*, 128: 33–36.

Jacob, M. J. and Grabner, M. (eds) (2010) *The Studio Reader: On the Space of Artists*. Chicago: University of Chicago Press.

Lawrence, T. B. and Dover, G. (2015) Place and Institutional Work: Creating Housing for the Hard-to-house. *Administrative Science Quarterly*, 60(3): 371–410.

Markusen, A. (2013) Artists Work Everywhere. *Work and Occupations*, 40(4): 481–495.

Massey, D. (2013) *Space, Place and Gender*. Cambridge: Polity Press.

Massumi, B. (2002) *Parables for the Virtual: Movement, Affect, Sensation*. Durham, NC: Duke University Press.

Michels, C. and Steyaert, C. (2017) By Accident or by Design. Composing Affective Atmospheres in an Urban Art Intervention. *Organization*, 24(1): 79–104.

Pullen, A., Rhodes, C. and Thanem, T. (2017) Affective Politics in Gendered Organizations: Affirmative Notes on Becoming-woman. *Organization*, 24(1): 105–123.

Ringelberg, K. (2017) *Redefining Gender in American Impressionist Studio Paintings: Work Place/Domestic Space*. Abingdon: Routledge.

Ruppel, S. (2020) Houseplants and the Invention of Indoor Gardening. In: J. Eibach and M. Lanzinger (eds) *The Routledge History of the Domestic Sphere in Europe*. London: Routledge, pp 509–523.

Schiebinger, L. (2007) *Plants and Empire: Colonial Bioprospecting in the Atlantic World*. Cambridge, MA: Harvard University Press.

Schilo, A. (ed) (2016) *Visual Arts Practice and Affect: Place, Materiality and Embodied Knowing*. London: Rowman and Littlefield.

Sedgwick, E. K. (2003) *Touching Feeling: Affect, Pedagogy, Performativity*. Durham, NC: Duke University Press.

Seigworth, G. J. and Gregg, M (2010) An Inventory of Shimmers. In: M. Gregg and G. J. Seigworth (eds) *The Affect Theory Reader*. Durham, NC: Duker University Press, pp 1–28.

Simpson, R., Morgan, R., Lewis, P. and Rumens, N. (2021) Living and Working on the Edge: Place, Precarity and Experiences of Male Manual Workers in a UK Seaside Town. *Population, Space and Place*, 27(8): 1–13.

Simpson, R., Morgan, R., Lewis, P. and Rumens, N. (2022) Landscape and Work: Placing the Experiences of Male Manual Workers in a UK Seaside Town. *Sociology*, 56(5): 839–858.

Swinth, K. (2001) *Painting Professionals: Women Artists and the Development of Modern American Art, 1870–1930*. Chapel Hill: University of North Carolina Press.

Taylor, S. (2015) A New Mystique? Working for Yourself in the Neoliberal Economy. *The Sociological Review*, 63(1): 174–187.

Taylor, S. and Littleton, K. (2016) *Contemporary Identities of Creativity and Creative Work*. London: Routledge.

Thomas, Z. (2015) At Home with the Women's Guild of Arts: Gender and Professional Identity in London Studios, c. 1880–1925. *Women's History Review*, 24(6): 938–964.

Thomson, M. and Borozdin-Bidnell, M. (2019) *Georgian and Regency Conservatories: History, Design and Conservation*. Liverpool: Liverpool University Press.

Thrift, N. (2004) Intensities of Feeling: Towards a Spatial Politics of Affect. *Geografiska Annaler: Series B, Human Geography*, 86(1): 57–78.

Thrift, N. (2007) *Non-representational Theory: Space, Politics, Affect*. London: Routledge.

Tyler, M. and Cohen, L. (2010) Spaces that Matter: Gender Performativity and Organizational Space. *Organization Studies*, 31(2): 175–198.

Waismel-Manor, R., Wasserman, V. and Shamir-Balderman, O. (2021) No Room of Her Own: Married Couples' Negotiation of Workspace at Home during COVID-19. *Sex Roles*, 85(11–12), 636–649.

Wapshott, R. and Mallett, O. (2012) The Spatial Implications of Homeworking: A Lefebvrian Approach to the Rewards and Challenges of Home-based Work. *Organization*, 19(1): 63–79.

2

Placing Postfeminism and Affect: Exploring the Affective Constitution of Postfeminist Subjectivities by Leaders in the City of London

Patricia Lewis

Introduction

This chapter brings together notions of affect and postfeminism to explore the affective positivity and affective dissonance connected to the identity work (Chowdhury and Gibson, 2019) of leaders working in the place of the insurance market in the City of London. While leadership as an activity has conventionally been characterized as a realm saturated by masculine norms, a social transformation in our understanding of the 'good' leader means that culturally feminine relational behaviours are increasingly expected alongside masculine-marked practices (Ford, 2006; Eagly et al, 2014). Consequently, a 'gender-balance' of masculine and feminine behaviours is seen as necessary for effective leadership (Gartzia and van Engen, 2012; Powell et al, 2021; Khan et al, 2022).

Postfeminism has produced the cultural conditions which facilitate the reframing of leadership as contradictorily gendered, and the reconfiguration of subjectivities characterized by the simultaneous uptake of masculine and feminine norms is fundamental to this discursive formation (Gill, 2007; Lewis, 2014, 2018; Chowdhury and Gibson, 2019). Interpellating individuals to engage in the constant mobilization of opposite gender norms, postfeminism calls for permanent movement 'back and forth between a grammar of hardness and a grammar of care', with the calibration between the two poles left up to an individual's 'own tactical calculations of intuition' (Brockling, 2005: 22). Significant research attention has been directed at the way postfeminism interpellates women in organizational contexts to connect with masculine norms through the enactment of empowered, entrepreneurial, agentic behaviours while simultaneously engaging in feminine displays of care, empathy and the nurture of others. In contrast, men's engagement with this discursive formation in terms of the calibration of masculine and feminine norms in the constitution of their subjectivities has had less research attention directed its way, despite recognition of an interdependence between the formation and shaping of postfeminist masculinities and femininities (Rumens, 2017).

However, shaping oneself in postfeminist terms – regardless of gender – is not only achieved through a disciplined response to discourses but is also influenced by affective attachments. As an affective phenomenon, postfeminism can be used as an analytic device, to make visible the emotions of leaders when engaged in the calibration of masculine and feminine norms. It can draw out how issues of leadership are navigated, and how leaders affectively respond to pressures generated by the continuous calibration of gendered discourses and norms. In doing this, a consideration of context is also important and I draw on the notion of place to highlight how the social and cultural are geographically constituted (Simpson et al, 2021). In presenting this analysis, the chapter explores how men as leaders manage the calibration of masculine and feminine norms and behaviours, highlighting the combination of affective dissonance and affective positivity which surrounds their gendered response to this calibration when engaging in leadership activities. In doing this, it reveals how gender is a felt force when moving between masculine and feminine norms within the place of the insurance market in the City of London. The chapter is organized as follows: first, the cultural phenomenon of postfeminism and how it is used as an analytic framework is outlined. I then present the methodology adopted for the study. This is followed by presentation of the empirical data and interpretation of the findings. The chapter finishes with a conclusion to the study.

Postfeminism and place

As both an object of analysis in and of itself and a critical concept drawn on to interpret contemporary gender relations, postfeminism has been intensively scrutinized, with much disagreement over how it should be interpreted and its value as a theoretical concept (Gill, 2007, 2016; Lewis, 2014, 2018; Gill et al, 2017). Within the gender and organization studies field, the prevailing understanding of postfeminism is one which treats it as a discursive formation connected to a complex set of discourses around gender, feminism and femininity, where equality is treated as part of our contemporary common sense and the ending of gender discrimination is emphasized (Lewis, 2014). Within this context, both men and women are called away from traditional gender norms, towards a reconfigured subjecthood, but this does not mean the obliteration of conventional masculine and feminine principles. Instead, what emerges is a coexistence of gender-equality norms and conventional gender norms discernible in reshaped modes of masculinity and femininity (Lewis and Benschop, 2023). This configuration shows as the combination of masculine and feminine behaviours such that together they form the central core of postfeminist subjectivities associated with discourses of individualism, choice and empowerment; self-optimization and constant transformation; 'natural' sexual difference; femininity as a bodily and psychological property; and active engagement with the care of children as a positive choice (Gill, 2007; Negra, 2009).

The interpellation to invest in a postfeminist subjectivity is conventionally understood in terms of an array of techniques which call individuals to constitute a (leadership) identity characterized by empowerment, choice and a hyper-engagement with the needs of others (Negra, 2009). These techniques do not emerge from a centrally located source but from the actions and influence of a range of semi-autonomous institutions, organizations and agents who deliver a similar message across public discourse, usually without organized cooperation (Riley et al, 2019). Given this, as both object and concept, postfeminism is constantly evolving, and consequently increasing consideration is given to the shifts *within* postfeminism and the sense that something has altered (Dejmanee, 2016). For example, postfeminism is now understood to have moved from a repudiation of feminism to the rehabilitation of a moderate feminism (Lewis et al, 2019; Banet-Weiser et al, 2020). As part of the examination of the malleability and tenacity of postfeminism, it is also now recognized as an affective and psychic phenomenon which tries to shape what women and men think and feel and how their emotional states are displayed (Gill, 2017). Accordingly, shaping oneself in postfeminist terms is not achieved solely through the disciplined take-up and performance of a set of life regulations. It is also dependent on affective attachments connected to a collection of 'feeling rules' which shape the way individuals articulate and manage the demands and contradictions of a (leadership) identity (Dobson and Kanai, 2019). In considering feeling in relation to the postfeminist constitution of leadership, I explore affect and discourse together. Here, I differ from the dominant approach that has turned away from discourse and meaning-making and suggests that there is division between thinking and feeling (Calder-Dawe and Martinussen, 2022). As such, the approach taken follows Wetherell's (2012) stance that affect and discourse cannot be separated and that emotional struggles take the form of affective discursive practices which are patterned within the everyday activities of place and social life.

In exploring the postfeminist calibration of masculine and feminine norms when constituting a leadership identity within the insurance market of the City of London, the latter, as said earlier, is more than just a background setting to the action of leading. Understanding the insurance market of the City of London in terms of 'place' means we can approach it not only as a location in the world but as a way of knowing that world (Cresswell, 2015). As one women leader in the wider study said about the insurance market:

'it's a very, very unique market, there has been a huge amount of jumping companies and that's how people build their network ... it's easy to jump, nobody has to change their commute ... so it's easy to jump ship around here and that lends itself to the village atmosphere, everybody knows each other because they've crossed paths somewhere along the line'.

Thus, the insurance market is an expression of a specific combination of geographical and social relations that includes the embodied relationship individuals have with the world (Simpson et al, 2021). A central social relationship within the place of the insurance market is that of gender, which has acted as a key reference point for how to 'do business'. In this regard, McDowell's (1997) study is seminal, as she drew out the entrenched masculinities that emanated from every aperture of the City of London, manifesting in the way business, trading, work and personal relationships were conducted. Nash (2018: 605), in a more recent study, sought to investigate how gender is 'currently situated in the City' in light of the entrance of women into leadership positions in growing numbers, focusing on 'what is experienced as in place and out of place'. Similarly, this chapter brings the concepts of postfeminism, affect and place together to explore how the changing relationship between masculinity and femininity in relation to leadership is 'felt' in place, and how the particularities of that situation are central to understanding how men calibrate masculine and feminine gender norms in the constitution of their leadership identity.

Methodology

This chapter stems from a project which sought to investigate how men and women leaders discursively constitute a leadership identity within the economic context of the City of London and the cultural location of postfeminism. A key aim of the study was to make visible the way in which postfeminist gender norms informed men's and women's leadership and the constitution of their leadership identities. With this purpose in mind, interviews were conducted with mid to senior leaders working in the insurance market within the City of London. A snowball sample was constructed from a mixture of personal and professional contacts and yielded 48 interviews. In this chapter the focus is on the 22 men who were interviewed as part of the study. Topics discussed during the interviews included the path to leadership, their motivation to lead, the work they did on themselves as they moved into leadership positions, their own account of how they lead and how this has changed over time, their views on what counts as 'good' and 'bad' leadership and the range of challenges they have faced over the course of their leadership career. The questions asked related to leadership experiences in general and were not specifically related to gender. This was to provide as open a space as possible for individuals to speak about leadership. However, the respondents were aware that men and women leaders were being interviewed and it was not unusual for the conversation to refer to gender in relation to leadership.

As the object of study was the postfeminist shaping of leadership identities within the insurance market of the City of London, a Foucauldian discursive methodology was adopted for analysis of the data. Discourse analysis techniques were used to investigate the way in which leadership identities were discursively constituted by respondents within postfeminist discourses, problematizing the common-sense association of men leaders with masculine behaviours and women leaders with feminized ways of leading (Lewis and Benschop, 2023). In taking this analytic approach, discourse is understood in Weedon's (1987: 12) terms as systems of text, concepts, beliefs and signs that exist in written and oral form and in the social practices of everyday life. Subject positions are offered by discourses, and their take-up by individuals is a discursive practice, activated by individual agency, but people are 'subjected to the power and regulation of discourse' (Weedon, 1987: 119). Three stages of analysis were undertaken: firstly, the focus was on the 'things said' by respondents in relation to their leadership activity as they invoked certain norms and subject positions 'establishing ways for people to be' (Bacchi and Bonham, 2016: 115). Secondly, the words and phrases articulated by interviewees were traced to postfeminist discourses as the utterances of the interviewees were located in and made sayable by them (Tassabehji et al, 2021). In other words, the 'things said' were read through the analytical lens of postfeminism and located within postfeminist constitutive discourses of individualism, choice and empowerment, makeover and self-transformation, care and relatability for those you lead and choosing active care for children and home. Discourses of individualism, choice and empowerment and makeover and self-transformation are understood to be connected to masculine attainment in work, while discourses of care and relatability for those you lead and active care for children and home are associated with feminized relational behaviour of building connection, focusing on the needs of others and providing support (Lewis, 2014; Lewis and Benschop, 2023).

One notable aspect of the interviews identified during data analysis was the emotional tenor of the discussion and the way positive and negative affect was generated in the interviewees as they sought to discursively constitute their leadership identity. As the strength of feeling articulated by the respondents was striking, this became a central focus of the analysis. Accordingly, focusing on the affect generated in the interviews and following Wetherell's (2015) argument that discursive practices are entangled with affect, the third stage of the data analysis concentrated on identifying the affective discursive practices that drew out how discourses are practised, circulated and felt within a culture. While there is often minimal description of emotion episodes within interviews, what is said is unquestionably 'felt'. Accordingly, paying attention to affective discursive practices 'dares discursively minded identity researchers to broaden their understanding of what counts as data and what matters for identity work' (Calder-Dawe and Martinussen, 2022: 129).

Affective discursive practices carve deep ruts in the bodies, brains, minds and subjectivities of individuals. Such grooves are a manifestation of the entanglement of the personal and social, as these undulations simultaneously impact on 'the social and cultural resources patterning and justifying social action' that is taken up and made personal by individuals as their own (Wetherell, 2015: 88). Affective discursive practices can reproduce accepted social norms or have a transformative potential as subject formation opens up opportunities for change in a subject position such as 'leader' (Lewis and Benschop, 2023). Adopting a focus on affect through the delineation of affective discursive practices aligns with the 'psy' shift in postfeminism, which, as said earlier, is now recognized as an affective and psychic phenomenon. Additionally, affective discursive practices locate the material world, such as places, within the horizon of identity research and wider social formations (Calder-Dawe and Martinussen, 2022). Within the analysis, consideration of the struggles and successes respondents expressed when calibrating masculine and feminine norms is presented as a set of affectively laden dilemmas. The next section is organized around the identified affective discursive practices and associated dilemmas connected to the issue of who to be as a leader and include disavowal of excessive individualism, valorization of the masculine, seeking and caring for aspiration and acceptance and regret around family. Excerpts from individual interviews that are illustrative of the identified affective discursive practices and resonate across the wider data set are included in the findings.

Findings

Disavowal of excessive individualism

Should care for others without caring too much

Postfeminism is strongly associated with individualism and the self-reliant individual who seeks internal, 'psy solutions' to challenges, pressures and barriers that emanate from outside themselves. There is a postfeminist cultural imperative to express and build an individualized potential by developing knowledge of an individual's interiority to support the call for constant work on the self. However, the discursive status of individuality and self-transformation is masculine, as it has always been open to men but only recently to women as they could not always access the terms of individuality (Cronin, 2000). However, the shifting terrain of individuality in terms of who can or cannot 'be an individual' has, within work contexts such as the insurance market, contributed to questions over levels of individuality. In narrating their leadership experiences, interviewees identified careerist, self-focused individuals as poor leaders whose behaviour has the potential to be detrimental to the business. Too strong a focus on a masculinized individualism without calibration with more feminized care for both business and people was seen by all respondents as poor leadership. An affective dissonance — feelings of being used, not being cared for, a sense of unpleasantness — was experienced by those subjected to the self-focused, individualistic leader, who was rejected as an example of poor leadership by respondents as follows:

'Here we had it [poor leadership] for a while, that was around self-interest when you know the reason they want to lead is because they want the profile and it is a stepping-stone to the next place, not about the current. And I've seen that a couple of times and it's very visible and it's just an unpleasant environment to work in when you know you're just being used and you know you've got someone who doesn't care, is making decisions around a perceived career progression rather than a real care for the business and the people.' (Interview no 3)

Nevertheless, while emphasizing the importance of care, collaboration and nurture in contrast to the strongly individualistic leader who is forging forward towards the future he can see, an affective dilemma around how much leaders should care emerges as follows:

'My wife says "(Name), why do you care about it (the organization) so much?" and I said, "Do you know what, I don't know". I feel a sense of obligation and duty and a sense that, you know, it's like a mission and I think this is ridiculous, this is an innate, you know, it's a building with a load of people in it and here I am, you know, worrying about the future of all these people and their families. So, so we want people, that's what we're looking for, is people who really care about it … and the organization is the people really, so I care a lot about the people.' (Interview no 20)

In expressing this view, this leader articulates his concern for the organization he leads through a postfeminist calibration of masculine and feminine norms within the place of the insurance market. Drawing on masculinized notions of duty, mission and obligation alongside a feminized sense of care for people and their families, an affective angst and feelings of some torment emerge from his narrative – he cares, but does he care too much? The worry expressed around care can be interpreted as emanating from the place of the insurance market, which 'comprises a collection of resources and conditions of engagement with the world' (Simpson et al, 2021: 8). This place is a gendered masculine space which impacts on the emotional responses articulated around the calibration of masculine and feminine norms when constituting leadership identity. Masculinity is embedded in the material elements of the insurance market, with men's bodies being highly visible in this place which is constituted around the strong masculine symbolic imagery of its phallic buildings (Nash, 2018). I suggest that the strength and feeling of masculinity that the insurance market projects acts to create a feeling of anguish in relation to how this can be calibrated with what is perceived to be the soft, feminist skill of care. Men within the place of the market are captured and constrained in terms of what they can say and do about their leadership, a constraint which is present throughout the data.

Valorization of the masculine
The importance of the feminine but the necessity of the masculine

The role of place in creating and maintaining the dominance of masculinity also emerges through the next affective discursive practice of the valorization of the masculine. The circulation of postfeminist discourses around the place of the insurance market has contributed to a growing value being attached to feminine leadership practices such as care and nurture of those you lead, but this has not usurped the value attached to masculine ways of leading, as the following illustrates:

> 'I think I'm extremely accessible that people can just walk in and just talk to me or talk to me while I'm walking around ... So, at the very top, my style wouldn't work at the very top, absolutely, we'd have anarchy and I don't think you can have that. So, I think you do need a degree of somebody who will force their will on to ... shape the organization and somebody in the organization needs to do that and I think those people who can do it, they're hugely impressive individuals ... (Name) ... is a hugely impressive individual, enormous achievements, his sheer force of will and character gets things done. He can be a nasty, unpleasant individual, really unforgiving and demanding ... without those traits you can't have that leadership style that he's got.' (Interview no 17)

Drawing on discourses of relatability and care and placing an emphasis on the importance of soft skills such as openness and a willingness to listen, this leader describes his identity as a leader in terms of a feminized collaborative leadership style. This is emblematic of the importance attached to soft skills by all respondents citing the need to 'bring out the best in people' within their organizations. However, he strongly criticizes his own leadership by stating that if this was dominant at the top of the organization there would be 'anarchy', a term which communicates intense conflicted feelings about his way of leading. His affective orientation to his own feminized leadership practices means that he constitutes himself as 'outsider', separating himself from the top tiers of leadership. This leads to the affective dilemma of the importance of the feminine but the necessity of the masculine. From his perspective, organizations within the place of the insurance market cannot survive and thrive without the masculine leader 'who will force their will on to … the shape of the organization'. Feelings of admiration for the power and strength of the conventional, heroic, masculine leader emanate strongly from this text. So, what if the masculine leader is 'nasty' and 'unforgiving', they can 'get things done', and without them success will be elusive. In his depiction of this masculine leader, this interviewee highlights the centrality of this masculinity to the place of the market – everyone knows these characters and their 'enormous achievements', many have experienced their unpleasantness, but it is these masculine leaders who constitute the insurance market and have made it what it is – without them there would be chaos.

Seeking aspiration

Repudiating human fragility, embracing human potential

The masculine materiality of the place of the insurance market – the visible dominance of men, masculinity and phallic buildings – influences the type of care which leaders are able and willing to give to those they lead. Care is not just one thing and a clear differentiation is made between care of fragility and care of potential, as the following fragments from one respondent demonstrate:

'I mean, I think a lot of people want to manage people for some reason and I never understand it. I never understand why people, because managing and leading, as you know, are somewhat different. Managing's a pain in the backside generally, isn't it? It's dealing with problems and nitty-gritty, at times, I'd love to have no people, the managing of things, you know, I come in on whenever, Tuesday, someone, first thing that happens is someone bursts into tears on me, in effect saying this group of people is not working well enough and I've got, ah, that's the last thing I really want, can't I continue leading,

leading the organization to new … so, I think there's a difference isn't there between that?' (Interview no 5)

In this respondent's text, care for people (and their problems) is narrated with significant affective dissonance. The individual who 'bursts into tears' is presented as contravening the positive and confident affective displays associated with successful working within a postfeminist era (Orgad and Gill, 2022). The affective discord he feels in this situation emerges strongly in his plea 'can't I continue leading the organization' to new horizons? Why do I have to deal with this excruciating, annoying, uncomfortable situation summed up in the phrase 'pain in the backside'? There is a disavowal of care for people which requires having to deal with negative emotions as opposed to confident, positive accounts of how a problematic work situation can be addressed. The affective dilemma here relates to the type of care as a leader that this interviewee is willing to provide, as he constitutes his leadership identity around the dialectic coexistence of masculine and feminine norms. Indeed, supporting people through their 'problems' is consigned to 'management' and is constituted as not leadership at all.

However, from the same interviewee, we see a shift from affective dissonance to affective positivity when it comes to the issue of nurturing capability, as the following illustrates:

'So, I'm a great fan of trying to meet people's aspirations. Yeah, also hit their full potential. Most people do not work to their full potential and it's really disappointing to me that that happens. Some people don't appreciate what their full potential is and you're trying to bring that out of them because the company will do better, and the individuals will be happier if we can get people really contributing and working to their full potential within the organization. So, I really like that bit, so the people side to me is the most exciting bit and then seeing people develop.' (Interview no 5)

Here, we see a completely different affective orientation to a people situation which is focused on supporting individuals to achieve their full potential. Embedded in this account of nurturing and caring for the individuals he leads are a set of feeling rules characteristic of postfeminism. These include an emphasis on positivity, being upbeat and working on confidence to support reaching an individual's goals. For this leader, care is (re)produced within the interdependent postfeminist discourses of individualism, self-transformation, care and relatability. We can see the affective positivity which infuses this care when he says he is a 'fan' of 'meeting aspirations' and that the 'people side' and 'seeing people develop' is 'exciting'. This is in marked contrast to the angst articulated in addressing people's problems in the earlier fragment.

In calibrating masculine and feminine norms when leading, this leader seeks to draw together his strongly felt desire to lead into the future with a form of care which is not overly feminized through its focus on people's problems but adjusts to align with the masculine place of the insurance market through its focus on potential.

Family: acceptance and regret
I chose work but I missed out on children

The final affective discursive practice relates to the issue of the relationship between work and family. There was an awareness among all respondents of how long working hours had a negative impact on family life, but reflection on this was viewed through the lens of choice. In placing an emphasis on choice, the interviewees took up the discursive stance of the choosing individual subject who freely engaged in long working hours (Lewis, forthcoming), as the following illustrates:

> 'But you know it's been tough ... so, you're doing a lot of travel, you know, I was probably as CEO [chief executive officer] of this business having client dinners three times a week so you know, you were only home at best at weekends and a lot of weekends you're sort of travelling ... I'm earning more money than I ever thought I'd earn in my life ... I made that decision and I'm happy with it. But it has been, you know, you've got to be conscious it is tough on the family.' (Interview no 3)

In taking this discursive stance, this interviewee communicates a strong, masculinized attachment-to-work characteristic of the insurance market. Work comes first, before family, and is just what you do if you hold a senior leadership position. Within the place of the insurance market, regular entertainment of clients is an expected and normalized business practice, as is constant travel. The impact of this is strongly felt, first in terms of the pleasure expressed around the money he earns, which reinforces a sense that he made the right decision and he is 'happy with it'. Nevertheless, there is also a feeling of concern about the impact this has had on his wife and children. Moving from affective positivity to affective dissonance, a sense of regret was communicated across the interviews, with some respondents wishing they could have done things differently, as the following demonstrates:

> 'I mean I've had a fantastic career, you know, I've been, I've travelled the world, made great friends, you know, done loads of exciting things and I don't know, I mean I, you know if somebody said to me I'd do it all over again so I'd just spend more time at home that's all

so … I think I would say that you've got, I would, I would want to do, I would want to do not the work bit but the home bit differently.'
(Interview no 20)

Here, the affective dilemma associated with putting work first and missing out seeing your children grow is expressed by this respondent through a reflection on how his work has impacted on family. We can see clearly the affective positivity embedded in his discursive stance expressed through reference to the excitement, fun and thrills that have characterized his career. Yet, affective dissonance is present in his statement that if he could 'do it all over again' he wouldn't change his work experiences but would do the 'home bit differently'. Drawing on the combination of discourses of individualism and choosing to be involved in active childcare, these respondents express a strong affective attachment to work but are also interpellated in their reflections by the postfeminist call to be a productive and reproductive citizen.

Conclusion

Placing postfeminism within the insurance market of the city of London, this chapter has sought to make visible the affective positivity and affective dissonance which emerge as male leaders engage in the calibration of masculine and feminine norms and behaviour to constitute a leadership identity. Previous research has highlighted a reluctance to engage with feminine norms and behaviours by male leaders, or, if taken up, there is a disinclination to name them in gendered terms (Fondas, 1997). This failure to name is located within the hierarchical masculine/feminine binary, where the feminine is still degraded as 'lesser' and non-mainstream. Indeed, Stemple et al (2015) assert that men enacting masculine behaviours are still seen as more legitimate in the leader role. Nevertheless, the contemporary postfeminist valorization of the feminized practices of care, nurture and empathy in relation to leadership means that characteristics identified with these gender norms cannot continue to be ignored. Instead, a modified adoption of behaviours such as the nurturing of others, conventionally identified as feminine, occurs with followers who are fun, positive and upbeat such that their potential can be nurtured for the benefit of their careers and the organization, and are more valued than those who are negative, fragile and pessimistic. Identifying the affective discursive practices (Wetherell, 2012) that are present in the place of the insurance market of the City of London, the chapter has highlighted the combination of affective dissonance and affective positivity which surround male leaders' gendered response to the calibration of masculine and feminine norms when engaging in leadership activities. In adopting this approach, I have made visible the way in which feminine gendered norms, while central to contemporary leadership, are modified by male leaders, influenced by the prevailing masculinity of the place of the insurance market. Affective dilemmas around care and variations in affective orientation towards different modes of nurture are implicated in the reproduction and reinforcement of the masculinity that has long been characteristic of the insurance market in London. While circulating postfeminist discourses have opened up economics arenas like the City of London to women, the masculinity which infuses this place is reinforced through the postfeminist calibration of masculine and feminine gender norms and is deeply felt by those who are subject to it.

Acknowledgements

The research reported in this chapter was supported by a Leverhulme Research Fellowship RF-2018-295 Postfeminism in the City.

References

Bacchi, C. and Bonham, J. (2016) Poststructural Interview Analysis: Politicizing Personhood. In: C. Bacchi and S. Goodwin (eds) *Poststructural Policy Analysis*. New York: Palgrave Pivot, pp 113–123.

Banet-Weiser, S., Gill, R. and Rottenberg, C. (2020) Postfeminism, Popular Feminism and Neoliberal Feminism? Sarah Banet-Weiser, Rosalind Gill and Catherine Rottenberg in Conversation. *Feminist Theory*, 21(1): 3–24.

Brockling, U. (2005) Gendering the Enterprising Self: Subjectification Programs and Gender Differences in Guides to Success. *Distinktion: Scandinavian Journal of Social Theory*, 6(2): 7–25.

Calder-Dawe, O. and Martinussen, M. (2022) Researching Identities as Affective Discursive Practices. In: *The Cambridge Handbook of Identity*. Cambridge: Cambridge University Press, pp 120–143.

Chowdhury, N. and Gibson, K. (2019) This Is (still) a Man's World: Young Professional Women's Identity Struggles in Gendered Workplaces. *Feminism and Psychology*, 29(4): 475–493.

Cresswell, T. (2015) *Place: An Introduction*, Second Edition. Oxford: Wiley Blackwell.

Cronin, A. M. (2000) Consumerism and 'Compulsory Individuality': Women, Will and Potential. In: S. Ahmed, J. Kilby, C. Lury, M. McNeil and B. Skeggs (eds) *Transformation: Thinking Through Feminism*. London: Routledge, pp 273–287.

Dejmanee, T. (2016) Consumption in the City: The Turn to Interiority in Contemporary Postfeminist Television. *European Journal of Cultural Studies*, 19(2): 119–133.

Dobson, A. S. and Kanai, A. (2019) From 'can-do' Girls to Insecure and Angry: Affective Dissonances in Young Women's Post-recessional Media. *Feminist Media Studies*, 19(6): 771–786.

Eagly, A. H., Gartzia, L. U. and Carli, L. L. (2014) Female Advantage Revisited. In: S. Kumra, R. Simpson and R. J. Burke (eds) *The Oxford Handbook of Gender in Organizations*. Oxford: Oxford University Press.

Fondas, N. (1997) Feminization Unveiled: Management Qualities in Contemporary Writings. *Academy of Management Review*, 22(1): 282–294.

Ford, J. (2006) Discourses of Leadership: Gender, Identity and Contradiction in a UK Public Sector Organization. *Leadership*, 2(1): 77–99.

Gartzia, L. and van Engen, M. (2012) Are (Male) Leaders 'Feminine' Enough? *Gender in Management: An International Journal*, 27(5): 296–314.

Gill, R. (2007) Postfeminist Media Culture: Elements of a Sensibility. *European Journal of Cultural Studies*, 10(2): 147–166.

Gill, R. (2016) Post-postfeminism? New Feminist Visibilities in Postfeminist Times. *Feminist Media Studies*, 16(4): 610–630.

Gill, R. (2017) The Affective, Cultural and Psychic Life of Postfeminism: A Postfeminist Sensibility 10 Years On. *European Journal of Cultural Studies*, 20(6): 606–626.

Gill, R., Kelan, E. K. and Scharff, C. M. (2017) A Postfeminist Sensibility at Work. *Gender, Work and Organization*, 23(4): 226–244.

Khan, M. H., Williams, J. and French, E. (2022) Post-heroic Heroism: Embedded Masculinities in Media Framing of Australian Business Leadership. *Leadership*, 18(2): 298–327.

Lewis, P. (2014) Postfeminism, Femininities and Organization Studies: Exploring a New Agenda. *Organization Studies*, 35(12): 1845–1866.

Lewis, P. (2018) Postfeminism and Gendered (Im)Mobilities. In: P. Lewis, Y. Benschop and R. Simpson (eds) *Postfeminism and Organization*, pp 21–42. New York: Routledge.

Lewis, P. (forthcoming) Consent and Work: A Postfeminist Analysis of Women Leaders' Acceptance of Long Working Hours. In: R. Ryan-Flood and J. James-Hawkins (eds) *Consent: Gender Power and Subjectivity*. London: Routledge.

Lewis, P. and Benschop, Y. (2023) Gendered Hybridity in Leadership Identities: A Postfeminst Analysis. *Gender in Management: An International Journal*, 38(2): 166–182.

Lewis, P., Adamson, M., Biese, I. and Kelan, E. (2019) Exploring the Emergence of Moderate Feminism(s) in Contemporary Organizations. *Gender, Work and Organization*, 26(8): 1063–1072.

McDowell, L. (1997) *Capital Culture: Money, Sex and Power at Work*. Oxford: Blackwell.

Nash, L. (2018) Gendered Places: Place, Performativity and Flânerie in the City of London. *Gender, Work and Organization*, 25(6): 601–620.

Negra, D. (2009) *What a Girl Wants? Fantasizing the Reclamation of Self in Postfeminism*. London: Routledge.

Orgad, S. and Gill, R. (2022) *Confidence Culture*. Durham, NC and London: Duke University Press.

Powell, G. N., Butterfield, D. A. and Jiang, X. (2021) The 'Good Manager' over Five Decades: Towards an Androgynous Profile? *Gender in Management: An International Journal*, 36(6): 714–730.

Riley, S., Evans, A., Anderson, E. and Robson, M. (2019) The Gendered Nature of Self-help. *Feminism and Psychology*, 29(1): 3–18.

Rumens, N. (2017) Postfeminism, Men, Masculinities and Work: A Research Agenda for Gender and Organization Studies Scholars. *Gender, Work and Organization*, 23(4): 245–259.

Simpson, R., Morgan, R., Lewis, P. and Rumens, N. (2021) Living and Working on the Edge: 'Place Precarity' and the Experiences of Male Manual Workers in a UK Seaside Town. *Population, Space and Place*, 27(8): 1–13.

Stemple, C. R., Rigotti, T. and Mohr, G. (2015) Think Transformational Leadership, Think Female? *Leadership*, 11(3): 259–280.

Tassabehji, R., Harding, N., Lee, H. and Dominguez-Pery, C. (2021) From Female Computers to Male Computors: or Where There Are so Few Women Writing Algorithms and Developing Software. *Human Relations*, 74(8): 1296–1326.

Weedon, C. (1987) *Feminist Practice and Poststructuralist Theory*. Oxford: Blackwell.

Wetherell, M. (2012) *Affect and Emotion: A New Social Science Understanding*. London: Sage.

Wetherell, M. (2015) Tears, Bubbles and Disappointment – New Approaches for the Analysis of Affective Discursive Practices: A Commentary on Researching the Psychosocial. *Qualitative Research in Psychology*, 12(1): 83–90.

3

Who's Afraid of Virginia Woolf? Affective Responses to Space, Objects and Atmosphere in a Writer's House Museum

Jessica Horne

Introduction

'She walked', remarked one volunteer room guide as we commenced our walking interview down the 'perilous' steps leading to the garden at Monk's House, a National Trust house museum in East Sussex. That *she* was Virginia Woolf, who formerly lived in the house, and it was as though this volunteer room guide had known her. This chapter discusses volunteers' affective responses to space, objects and atmosphere in the context of a writer's house museum, and how they develop attachments to place, defined here as people's affective bonds with a place or setting (Rubstein and Parmelee, 1992). Consistent with the demographic of National Trust volunteers, most of the volunteers at Monk's House are middle-class women, many of whom are volunteering in retirement to 'ensure places [they] love are preserved for generations to come' (The National Trust, 2023). Some of the volunteers I interviewed have been working on an unpaid basis at the museum for nearly 30 years. In this chapter, I draw on data from my walking interviews with volunteer room guides to demonstrate their affective responses to the museum. To this end, I am guided by Sara Ahmed's (2010: 33) understanding that through affect 'we are moved by things. And in being moved, we make things.' In the case of Monk's House, I show how the volunteers are moved by the material environment, such as the gardens, rooms, objects and an atmosphere punctuated by the absence of Virginia and Leonard Woolf. I observe how the volunteers' affective responses shape their perceptions of the past inhabitants of the house and the spaces they would have used. I argue that in the process of being 'moved', the volunteers 'make' meanings for themselves and visitors (Ahmed, 2010). My findings suggest that for some, space, objects and atmosphere may be therapeutic, especially for those coping with some of the challenges typically associated with later life. For others, particularly LGBTQIA+ (lesbian, gay, bisexual, transgender, queer/questioning, intersex, asexual) people who orientate towards objects that are hidden away, affective responses to the past help them carve out spaces for belonging in the present.

Affect in organization studies

Affect can be thought of broadly as 'a body's capacity to affect and be affected by things' and that a body in this case may be human, non-human or part-human (Seigworth and Gregg, 2010: 2). Critical studies of affect are concerned with 'what is evoked by bodily experiences as they pass from person to person, in a way that is contagious but remains unspoken' (Fotaki et al, 2017: 4). According to Fotaki et al (2017), affect has been considered from four distinct perspectives in organization studies. They outline that a first strand of scholarship draws inspiration from psychoanalysis to understand the social implications of affective impulses or, put simply, the relational implications of affect, such as those explored by Judith Butler. A second strand of affect studies emerging from non-humanist traditions understands affect as 'a visceral force, or a vector, that operates between bodies' with a view to better understand relationality (Fotaki et al, 2017: 5). A third stand focuses on 'spatiality as redefining of subjectivity as an insatiable desire for freedom that can only be achieved through endless becoming' (Fotaki et al, 2017: 6). Lastly, a fourth strand explores the transformative power of affect by focusing on 'politics and resistance to normativizing power' (Fotaki et al, 2017: 6). I align myself with this fourth strand of scholarship, informed by queer and feminist theory, and in doing so I recognize the potentially transformative power of affect for people volunteering in house museums and the 'generative effects' for reproducing and challenging (hetero)normative power structures.

Feminist geographer Deborah Thien (2005: 451) draws a necessary distinction between affect and emotion, arguing that 'affect is the *how* of emotion' and is 'used to describe (in both the communicative and literal sense) the motion of emotion' (original emphasis). Writing from a queer and feminist perspective, Sara Ahmed (2010) invites us to consider the ways that emotions are produced through affect, suggesting that we orientate towards things that make us happy. For Ahmed, 'to be affected by something is to evaluate things. Evaluations are expressed in how bodies turn toward things' (2010: 31). Taking 'good feeling' as a starting point for consideration in this work, Ahmed suggests that even if the object of happiness is lost, the feeling of being affected by the object remains. Ahmed argues that we orientate ourselves towards or away from objects depending on how we are affected by them, such that 'to be affected "in a good way" involves an orientation toward something as being good' (2010: 32) or we move away from things that make us unhappy. In directing ourselves towards an object, we are 'aiming somewhere else' or, as Ahmed puts it, 'we aim for happiness, if happiness is what we get when we reach certain points' (2010: 34). This raises questions for me about how volunteers orientate themselves towards certain spaces, objects and atmospheres at Monk's House, to be affected 'in a good way' by their volunteering (Ahmed, 2010). In summary, then, this chapter draws on Sara Ahmed's (2010) theory of 'happy objects' to explore volunteers' affective responses to space, objects and atmosphere in a house museum, and the potentially transformative nature of these responses for themselves and visitors.

Affect in museums

Museums have been conceptualized as affective spaces, eliciting emotional responses from people who visit (Varutti, 2023). In museum studies, Andrea Witcomb (2013) explores the role of curatorial practices in producing affective experiences for visitors in museums. Witcomb (2013) makes the case that affective spaces can be used intentionally to provoke unsettlement among visitors, such that they are forced to consider the 'self' in relation to the 'other'. Drawing on examples from museums in Australia, she suggests that in some affective spaces, visitors are forced to ask questions about their own position in relation to the dominant narrative being offered by the museum. This speaks to the transformative potential of affect which museums can harness in the present by orchestrating engagement with the past (Fotaki et al, 2017). Historic house museums have their own particularities, having been formerly inhabited by people, museumized and presented still as dwellings (Young, 2015). Gregory and Witcomb (2007: 625) posit that house museums can facilitate 'affective engagement with the past'. They argue that the very palpable sense of absence in house museums means that 'historic houses open up a space in which the intangible past can be sensed', going on to add that 'in silence, in gaps' we may sense the 'presence' of the people that used to live in the house (Gregory and Witcomb, 2007: 265). However, they caution that in spaces where the 'intangible past' can be sensed, an 'atmosphere' marked by absence can 'frequently collapse into mute, static pictures of the past, which do not affectively speak in the present' (Gregory and Witcomb, 2007: 265).

In the examples provided, Witcomb considers visitors' affective engagement with the past. As Holmes and Edwards (2008) argue, volunteers in heritage attractions can be thought of as special type of visitors who gain a different experience of visiting through their volunteering. Unlike regular visitors, volunteers spend extended periods of time in the spaces of the museum through the special access afforded by their volunteering, and they enjoy close encounters with the museum's artefacts. In this chapter, I foreground volunteers' affective responses to the materiality of a house museum, such that it allows them to engage with the past and consider their sense of place and belonging in the present. Scholars have gone some way to explore the role of affect in heritage volunteering. In organization studies, Ward and Greene (2018) investigate the role of affect for volunteers in the National Trust. Arguing that the 'affective turn' has been largely overlooked in relation to non-profit and voluntary organizations, they found that volunteers in the National Trust demonstrate an 'affective commitment' to the space, place and stories of the properties where they volunteer, such that they may be resistant to changes imposed by paid staff (Ward and Greene, 2018). They argue that specific emotions such as 'love, affection, passion and pride' evidence volunteers' affective commitment to the museum (2018: 1164). In their case, the outcome of volunteers' affective commitment is discussed in relation to management outcomes; however, in this chapter I explore the outcomes of volunteers' affective responses to the space, place and atmosphere of the museum for themselves and their interactions with visitors.

Methods

This chapter draws on data from my research conducted in July 2019 with volunteers at Monk's House. At the time of my study, there were approximately 80 volunteers at the property and six paid members of staff. My aim was to explore the volunteers' relationship to place and the wider cause of the National Trust. Recognizing the 'complexity of heritage sites', Ciolfi and Petrelli (2015: 48) advocate for the use of walking as a method to understand the relationships that heritage volunteers have to the places of their volunteering. Following this, I developed a variant of the 'go-along' walking interview to be conducted with eight volunteers in the grounds of the museum. 'Go-along' interviews are helpful because they allow researchers to simultaneously interview and observe participants (Kunsenbach, 2003; Evans and Jones, 2011), and walking methods have been celebrated for offering insight into people's affective and bodily responses to place (Truman and Springgay, 2019). My interviews were semi-structured, and the questions invited the volunteers to reflect on why they chose to volunteer, what they do on a normal day as a volunteer, their relationship to space and place and whether this had changed over the course of their volunteering. Seven of my participants were room guides who volunteered in the house and one participant was a garden volunteer. All but one of my participants were white, middle-class women, reflective of the demographic of volunteers in National Trust properties (Harflett, 2015). Most of my participants were volunteering in retirement; however, one participant was a university student who was comparatively younger than the other volunteers. Some of the women in my study had been volunteering at Monk's House for nearly 30 years and had extensive involvement with the museum. In the findings that follow, I highlight the volunteers' affective responses to space, place and the past inhabitants of the museum, and demonstrate how, in being 'moved', volunteers 'make' meanings for themselves and visitors (Ahmed, 2010).

Volunteers' affective responses to space and objects

Historic house museums offer possibilities for affective encounters with the past in the present (Gregory and Witcomb, 2007). Writer's house museums, in particular, help us to imagine how authors may have lived and how this informed their work. During my walking interviews with volunteer room guides at Monk's House, it was clear that many of the volunteers had extensive knowledge of the property where they had spent extended periods safeguarding the collection and talking to visitors. It was no surprise to me when they began to reveal their affective responses to space, place and objects at the museum. I observed that some of their affective responses were shaped by the gendered division of space at the museum. To contextualize, Monk's House and the National Trust suggest to visitors that the bedroom and writing lodge are rooms that were used almost exclusively by Virginia Woolf. Many of the volunteers said that they enjoyed volunteering in the bedroom, describing the room as having a "distinct atmosphere" and a "nice feel to it". When I asked Bev how it feels to volunteer in that room, knowing that it was a space used exclusively by Virginia Woolf, she said that:

> 'I like to feel that she was happy there. I like people who come in to be able to go away thinking that she was really happy in that space, and not to concentrate on her mental state and going off and committing suicide, but that this was such a happy place to go from. Yes, I feel at peace in there. It's a very calming place. I think lots of people do think "this is a lovely room" and I like them to think that she was so happy at Monk's House, rather than sad.'

In this quote, Bev emphasizes that Woolf was 'happy' in the house, but particularly in her bedroom. Knowing that Virginia was happy in the bedroom, Bev is drawn to the room and feels 'at peace' when she is volunteering there. In being 'moved' by the room and feeling 'at peace' there, Bev helps to 'make' new understandings for visitors, who might have thought that Woolf was unhappy at Monk's House due to the circumstances surrounding her death (Ahmed, 2010). We see here that Bev's affective response informs her approach to her voluntary role and encounters with visitors in the bedroom. In her work, Sara Ahmed (2010: 114) suggests that our experience of an 'atmosphere' is shaped by the 'angle of our arrival' to that atmosphere, and we can also consider the atmosphere as angled and felt from a specific point. During my interviews, I observed that the volunteers' 'angle of arrival' to their volunteering shapes their affective responses to space and place at Monk's House. For example, Anna began volunteering in retirement at a time when her elderly mother's health was declining. Later, and at the time of my interview with Anna, she had just returned after a period of ill health which had threatened the continuation of her volunteering. We paused our walk in the garden to talk about Virginia Woolf, and Anna said:

'Well, [Monk's House] did keep her going didn't it, I mean that's almost why I can sort of relate to that, because she used to come down and everything else faded away and it was her complete place of tranquillity. As I say, it saved me in the first place and then it saved me again recently, in the last six months, because I had a goal to come back to. It makes you feel a bit more worthwhile, and if you're just messing around at home or fiddling about in the garden or cleaning the house or something, that's not a good thing when you've retired, you need something like this. But basically, it's kept me going, a bit like her.'

Anna clearly relates to Virginia Woolf's experience of having been 'kept going' by being at Monk's House. She is affected by the space in the same way that she imagines Virginia Woolf once was, and she can appreciate how 'everything faded away' when she was at Monk's House. I notice from Anna's description that she 'angles' the atmosphere at Monk's House to remind us of the sense of 'tranquillity' it provided Virginia Woolf. Anna's 'angle of arrival' to this atmosphere and her volunteering is one of needing something to sustain her through some of the challenges associated with later life, including, in her case, the illness of an elderly relative and the perceived loss of routine from paid work in retirement (Ahmed, 2010). Anna's affective response to Monk's House is enduring, as even after being away from the museum due to a period of illness she orientates herself back to this place which 'kept [her] going'. Volunteering is itself a place that Anna can return to, relatively unseen by a structure of the household and reproductive labour in retirement, and it makes her feel 'worthwhile'.

In these cases, the volunteers' affective responses are not only shaped by their 'angle of arrival' to Monk's House, but also by the narratives surrounding the gendered division of space at the museum. The volunteer room guides all stressed that the bedroom and writing lodge were very much 'Virginia's spaces' where she could be at 'peace'. Despite this, the volunteers recognized the centrality of other spaces at the property in producing affect, including the garden. Lydia, the only garden volunteer I interviewed, described herself as feeling "more attuned to Leonard" as "the garden was his area". She herself had 'arrived' at her volunteering in retirement and felt compelled to volunteer following the death of her husband. When I asked Lydia what attracted her to Monk's House, she said:

> 'I just fell in love with this place as soon as I walked around the corner. It seemed so peaceful and tranquil and that was what I was searching for, but also being able to do a hobby in a safe environment and with people, a group of like-minded people, and just it struck me as being a very healing type of place.'

Lydia's expression of her 'love' for this place is suggestive of the value of Monk's House at this stage of her life. Like others, Lydia appreciates the peace and tranquillity of the garden, particularly at a time when she is grieving the loss of her spouse. Lydia later attributed the distinct atmosphere of the garden to Leonard Woolf.

Many of the volunteer room guides I spoke to suggested that although Leonard 'sometimes gets overlooked' at Monk's House, he promoted a necessary stability that facilitated Virginia's writing and protected her mental health. Anna, for example, explained that "Leonard was so protective of her, and so caring", admitting that "she would never have lasted as long without him". She and others appreciated the 'tranquillity' of the garden that they attributed to Leonard's careful design of the garden into four contrasting sections, using the remains of former cowsheds as the framework of the garden (The National Trust, 1998: 15). For example, Shelley led me to one area of the garden and said that "if I'm sat in the garden, I like to sit by the fishpond here, in the middle. It's just the peace and tranquillity here, it's beautiful, it's really nice." Through their affective responses to the bedroom and garden, both gendered spaces, heteronormativity is made and remade by some of the volunteers at the museum. Scholars in museum studies have discussed the heteronormativity of house museums, including National Trust properties (see, for example, Smith, 2017; Curran, 2019), acknowledging the attempts made by organizations like the National Trust to foreground the LGBTQIA+ histories of their properties (see Sandell et al, 2018). In her work, Alison Oram (2012) suggests historic house museums are 'dynamic sites' with multiple layers of interpretation and that the spatial arrangement of museums contributes to the circulation of discourses, including those that make normative assumptions about the sexuality of the house's past inhabitants (Oram, 2012: 537). Here I show that in being 'moved' by the peaceful atmosphere in the garden, a space attributed to Leonard, the volunteers contribute to the normativization of heterosexuality at the museum through the reproduction of gendered spaces.

While some talked more about their attachment to rooms, and space and place more generally, other volunteers described their attachments to smaller objects and items of the collection. For university student Angelica, volunteering for only a few weeks in the summer, she admitted that it is the "little things that are tucked away that people don't notice, and you can talk about" that are meaningful for her volunteering. Angelica directed my attention towards a small object, describing it as "a little blue ashtray that Vita Sackville West gave Virginia", explaining that "they had matching ones, it's cute, really cute". Angelica's orientation towards this object resonates with Sara Ahmed's (2010: 32) proposition that 'to be affected "in a good way" suggests an orientation to something as being good'. Here, Angelica orientates towards the ashtray because it is something that brings her happiness. She is *moved* by this little object, and 'in being moved' she makes way for an alternative discourse that acknowledges the significance of Woolf's love affair with Vita Sackville West (Ahmed, 2010: 33). In being affected by the little object, Angelica carves out a space within which she can question:

'Why are we fitting love into these confines that you know, society and heterosexuality have imposed on us already? And that somehow, queer people now are like "oh we have to be dating, we have to be girlfriends or this and that" and it's like, things are more complicated than that. Even Virginia and Leonard, they weren't really married in that sense.'

In this example, we see how Angelica's affective encounter with this little object destabilizes the (hetero)normative order maintained by some of the volunteers' responses to the gendered division of space at Monk's House. Through this tiny blue ashtray, Angelica not only acknowledges Woolf's lesbian affair but also queers her marriage to Leonard, recognizing that the couple contested some of the norms of heterosexuality. The point I wish to make is that volunteer room guides like Angelica may turn to corners, things that are 'tucked away', to be moved themselves, and to make new meanings for visitors. In this case, Angelica moves towards an object through how it affects her, as it helps her carve out a sense of belonging and representation for herself as a queer woman that had to be searched for against the backdrop of the other volunteers' affective responses to gendered spaces. Queer is not only a sexual orientation but, as Sara Ahmed (2006) argues, queerness involves 'disorientation' to conventional norms, creating space for thinking and inhabiting the world differently. Here, Angelica's perspective brings forth the centrality of Virginia Woolf's relationship with Vita Sackville West (who famously inspired her novel *Orlando*), challenging those who reduce their relationship to the line 'they were very close friends'. In being affected by the object, Angelica helps us consider queer perspectives on the past that are not often foregrounded in the dominant narratives of the museum.

What I would like you to take away from these examples is that for the women I spoke to, many of whom are volunteering later in life and in retirement, affective responses to space and place help to shape their lives in the present. Some of the volunteers orientate themselves towards volunteering at Monk's House because it brings them happiness and helps them feel 'at peace'. These accounts mirror the findings of Ciolfi and Petrelli (2015: 50), who acknowledge that many people choose to volunteer at heritage sites for the 'peacefulness' and 'therapeutic atmosphere' that they can afford. Importantly, I have shown that the volunteers' affective responses to the house and garden inform the (re)production of gendered spaces at Monk's House, such that heteronormativity is preserved in narratives that are circulated by the museum. While some describe their affective responses to spaces and rooms, others are moved by the less obvious objects in the house. In affective responses to objects that are tucked away, the transformative potential of affect is realized for some queer people, searching for belonging in the present through engagement with the past.

Conclusion

This chapter has explored volunteers' affective responses to space, objects and atmosphere in the context of a writer's house museum. Drawing on Sara Ahmed's (2010) conceptualization of 'happy objects' which we orientate towards depending on how they make us feel, this chapter has shed light on the role of affect in volunteers' attachments to space and place. The volunteers I interviewed spend extended lengths of time in the house and garden through their volunteering and respond affectively to the house's past inhabitants and the spaces they would have used. In being affected by the materiality of the house, including spaces, objects and atmosphere, I argue that volunteers (re)produce gendered spaces at the museum, resulting in a (hetero)normative perspective on the past being communicated. I have demonstrated that in being affected by the museum, the volunteers make meaning for themselves and visitors in the present. I have illustrated that particularly for women volunteering in retirement, the transformative power of affect lies in the fact that affective bonds to place are therapeutic, providing some people with a sense of purpose during changes in later life (Fotaki et al, 2017). I wish to argue that through their affective responses, the volunteers welcome the peace and tranquillity that volunteering provides. I agree with Gregory and Witcomb (2007) that affect does not serve all in the present when it leads to nostalgia and a romanticized view of the past. As Joshua Adair writes, 'it is essential to acknowledge that lesbians are as woefully under-represented in house museums as gay men' (2010: 267). The marked absence of Virginia and Leonard forces the volunteers to comprehend their past lives at Monk's House in the present, and in being affected by an 'atmosphere' punctuated by their absence, they are 'moved' and 'make' interpretations of their lives for visitors in the present. In the peace and tranquillity, they find some volunteers do not disrupt the gendered spaces at the property such that some volunteers seek out smaller, hidden objects to make visible LGBTQIA+ perspectives on the past that offer them a sense of belonging in the present. And considering Sara Ahmed's (2023: 220) feminist killjoy commitment of being 'unwilling to make happiness my cause', I challenge readers to consider how volunteers' orientations to happiness do little to challenge normative representations of the past, including in house museums that fail to address their colonial legacies.

References

Adair, J. (2010) House Museums or Walk-In Closets? The (Non)Representation of Gay Men in the Museums they Called Home. In: A. K. Levin (ed) *Gender, Sexuality and Museums: A Routledge Reader*. New York: Routledge, pp 264–278.

Ahmed, S. (2006) Orientations: Toward a Queer Phenomenology. *GLQ: A Journal of Lesbian and Gay Studies*, 12(4): 543–574.

Ahmed, S. (2010) Happy Objects. In: M. Gregg and G. Seigworth (eds) *The Affect Theory Reader*. Durham, NC: Duke University Press, pp 29–51.

Ahmed, S. (2023) *The Feminist Killjoy Handbook*. Milton Keynes: Adam Lane.

Ciolfi, L. and Petrelli, D. (2015) Walking and Designing with Cultural Heritage Volunteers. *Interactions*, 23(1): 46–51.

Curran, S. (2019) Queer Activism Begins at Home: Situating LGBTQ Voices in National Trust Historic Houses. Doctoral dissertation, University College London.

Evans, J. and Jones, P. (2011) The Walking Interview: Methodology, Mobility and Place. *Applied Geography*, 31(2): 849–858.

Fotaki, M., Kenny, K. and Vachhani, S. J. (2017) Thinking Critically about Affect in Organization Studies: Why it Matters. *Organization*, 24(1): 3–17.

Gregory, K. and Witcomb, A. (2007) Beyond nostalgia: The role of affect in generating historical understanding at heritage sites. In: S. Knell, S. MacLeod and S. Watson (eds) *Museum Revolutions: How Museums Change and Are Changed*. New York: Routledge, pp 263–275.

Harflett, N. (2015) 'Bringing them with Personal Interests': the Role of Cultural Capital in Explaining Who Volunteers. *Voluntary Sector Review*, 6(1): 3–19.

Hemmings, C. (2005) Invoking Affect: Cultural Theory and the Ontological Turn. *Cultural Studies*, 19(5): 548–567.

Holmes, K. and Edwards, D. (2008) Volunteers as Hosts and Guests in Museums. In: K. D. Lyons and S. Wearing (eds) *Journeys of Discovery in Volunteer Tourism: International Case Study Perspectives*. CABI, pp 155–165.

Kusenbach, M. (2003) Street Phenomenology: The Go-along as Ethnographic Research Tool. *Ethnography*, 4(3): 455–485.

Oram, A. (2012) Sexuality in Heterotopia: Time, Space and Love between Women in the Historic House. *Women's History Review*, 21(4): 533–551

Rubinstein, R. L and Parmelee, A. (1992) Attachment to Place and the Representation of the Life Course by the Elderly. In: I. Altman and S.M. Low (eds) *Place Attachment, A Conceptual Inquiry*. New York: Plenum Press, pp 139–160.

Sandell, R., Lennon, R., Smith, M. and Lincoln, A. (2018) *Prejudice and Pride: LGBTQ Heritage and Its Contemporary Implications*. University of Leicester. Report. https://hdl.handle.net/2381/42909

Seigworth, G. and Gregg, M. (2010) An Inventory of Shimmers. In: M. Gregg and G. Seigworth (eds) *The Affect Theory Reader*. Durham, NC: Duke University Press, pp 1–25.

Smith, M. (2017) Queering the Historic House: Destabilizing Heteronormativity in the National Trust. In: B. Pilkey, R. Scicluna, B. Campkin and B. Penner (eds) *Sexuality and Gender at Home*. London: Routledge, pp 105–120.

The National Trust (1998) *Virginia Woolf and Monk's House*. Swindon: Acorn Press Ltd.

The National Trust (2023) *Volunteer with us*. Available at: www.nation altrust.org.uk/support-us/volunteer/volunteer-with-us (accessed 15 June 2023).

Thien, D. (2005) After or beyond Feeling? A Consideration of Affect and Emotion in Geography. *Area*, 37(4): 450–454.

Truman, S. E. and Springgay, S. (2019) Queer Walking Tours and the Affective Contours of Place. *Cultural Geographies*, 26(4): 527–534.

Varutti, M. (2023) The Affective Turn in Museums and the Rise of Affective Curatorship. *Museum Management and Curatorship*, 38(1): 61–75.

Ward, J. and Greene, A. M. (2018) Too Much of a Good Thing? The Emotional Challenges of Managing Affectively Committed Volunteers. *Nonprofit and Voluntary Sector Quarterly*, 47(6): 1155–1177.

Witcomb, A. (2013) Understanding the Role of Affect in Producing a Critical Pedagogy for History Museums. *Museum Management and Curatorship*, 28(3): 255–271.

Young, L. (2015) Literature, Museums, and National Identity; or, Why Are there so Many Writers' House Museums in Britain? *Museum History Journal*, 8(2): 229–246.

4

Trading from Home: The Affective Relations of 'Doing Finance' in the Domestic Setting of the Home

Corina Sheerin and Alex Simpson

Introduction

The coronavirus pandemic and the resultant response of nationwide or local lockdowns has had a seismic impact on labour markets, triggering new relationships with our work environments. While much attention has since been placed on the 'new normal' and the legacy the pandemic is leaving in relation to work practices, remote work, precarity and work hours across multiple sectors (Collins et al, 2021; Vyas, 2022), for the financial services industry in the City of London the impact was perhaps more fleeting but not less important (Tooze, 2021). For around ten months across 2020 and 2021, the UK was in lockdown, forcing nearly all the City of London's market-making financial operations to be emptied from their slick, glistening, high-rise towers to the more mundane and domestic setting of the home. The usual throng of Leadenhall Street or the frantic whir of Cornhill became muted for the first time since the World War II, as finance workers plugged their Bloomberg controls and multiple screens into spare rooms and kitchen tables around the country. For two sociologists interested in finance, gender and work practices, this was a unique opportunity to examine the experiential shift from the aggressive, fast-paced masculine sphere of the City to the more domesticated sphere of the home.

As McDowell (1997: 12) argues, the physical construction of the workplace affects, as well as reflects, the social construction of workplaces and their inhabitants within a relation to power. Just as the City is a social, cultural and economic site of privilege, so too are the finance workers who occupy the elite banking and financial institutions. This is to both place and understand bodies and their socially encoded meaning in relation to the specific spatial, temporal and cultural context in which they are placed; one which creates the conditions of the intergeneration transmission of shared gendered outlooks and values (Riach and Cutcher, 2014). In the City of London, as with other global financial centres, the spatial frame has served only to promote and reify a dominant vision of masculinist aggression, greed and competitive one-upmanship. For example, McDowell (2010; 1997) and Zaloom (2006) highlight the impact of the traditional open-outcry exchanges on performative financial relations. However, as a result of the COVID-19 pandemic, the City of London fell largely dormant, as its trading work force had been shifted from the masculine place of finance work to the feminized sphere of the home. This chapter capitalizes on this opportunity to offer a unique insight into the challenges, possibilities and experiences of conducting a masculinist occupational endeavour in the largely feminized sphere of the home, examining key themes of detachment, loss, tension and negotiation as well as opportunism and possibility. Through this temporary reconfiguration of spatial work relations, we reveal the ways in which hierarches, social relationships, organizational norms and power dynamics were upheld and reconfigured, focusing on the tensions of (gendered) identity formation within this new blurred landscape between the financial and domestic.

Gender, space and finance

When addressing questions of finance and space, it is always important to remember that, despite four decades of global market integration and major advances in electronic trading, and for all its ideological abstractions, finance continues to be rooted within – and shaped by – a close material, spatial and architectural make-up (Sassen, 2005; Simpson, 2018). The City of London, in other words, offers a local and human-scale spatial environment in which the abstractions of the financial markets are rooted (Longlands, 2020; Simpson, 2021). The City's architectural and topographical prominence looms large across London's skyscape, with its increasingly vertical hue overshading the historical networks of lanes and alleys that lie beneath. Standing below these smoothed façades and friction-free glass-and-steel edifices, it is possible to detect a distinct atmosphere that points to, as Spencer (2016) summarizes, an aesthetic of vastness, blankness, pointing, incongruity and intensive coherence. Yet, for the City workers who pass through the streets on a daily basis, the scale and grandeur of these buildings is a reminder of who delivers and is responsible for this collective wealth and success. Bringing a coherent structure of oneness, wholeness and greatness, these vast steel-and-glass structures cast, in the language of Wacquant and Bourdieu (1993), a 'great reservoir of symbolic power' to underline the power and ubiquity of the market mechanism. With this comes a sense of belonging that, for

Spencer (2016), legitimates a form of power that operates both on and through the constitution of the self. It is here, buried within this landscape of prominence, that finance workers 'weave back and forth between actual and digital space' (Sassen, 2005: 32) in the undertaking of their occupation. After all, finance workers are connected through technology, yet remain rooted within the spatial confines of trading and office environments, in a way that is emulated by few other sectors (Zaloom, 2006).

As a material tapestry of power, these layers of wealth and the spatial structure of finance are perpetually pressed and shaped in an orchestrated manner to establish a common framework of truth, justice and morality. It is also important not to overlook the constructions and expressions of gender that are woven into this landscape. After all, as Sylvia Walby (2015) reminds us, the project of neoliberalism is deeply gendered and, along with it, the construction and maintenance of finance. There is a need, therefore, to place and understand gender and its socially encoded meaning in relation to the specific spatial, temporal and cultural contexts of finance; one which creates the conditions of the intergenerational transmission of shared gendered outlooks and values (Riach and Cutcher, 2014). The senior realms of finance are – and have always been – dominated by men, especially the market-facing 'front office' roles of investment management, trading, sales and broking. Generally, such roles are knowledge intensive, well paid and highly sought after, while also characterized by 'the boys' cliché, which is woven into a spatial distribution that reinforces gender inequality (Sheerin and Linehan, 2018). For example, in its 2021 Alpha Female, CityWire indicated that of the 16,563 fund managers surveyed, only 11.8 per cent were female, while in the UK context, women represented 12 per cent of all fund managers, 11 per cent in Ireland and 9 per cent in Australia (CityWire, 2021: 4–6). Historically, women have been actively excluded from participating in financial trading and were barred from becoming members of the New York Stock Exchange until 1967, 150 years after its first opening. It took even longer for women to be admitted into the London Stock Exchange (LSE). The LSE was founded in 1698, but it was not until 1973 – 275 years later – that the first woman was allowed membership (Baeckström, 2022).

It is little wonder, then, that men have dominated – and continue to do so – in terms of numerical representation at senior levels, with a pervasive culture embodying a masculinist disposition where women are portrayed as 'not [a] natural fit' and located as 'outsiders' (McDowell, 2011: 175; Fisher and Kinsey, 2014). McDowell (2010: 653), in her study of culture within the City, revealed that exaggerated forms of masculinized language and behaviour with 'horseplay, sexualized banter … as well as forms of sexual harassment' are commonplace in investment banking and finance more broadly. Gender bias, power dynamics and inequality are part of the fabric of the sector. Such norms are socially constructed collective processes which are perpetuated through the neoliberal and masculinist formal and informal mechanisms which underlie the sector. Regimes such as a long hours culture, sexism, presenteeism, gender pay gaps, lack of access to networks, sponsors and role models as well as social capital reinforce the hegemony (Hertz, 1998; Ho, 2009). The very culture of greed, excess, hyper-aggressive and competitive endeavour, which permeate through and structure the world of finance, can each be distilled to the 'virility' that McDowell (1997) first recounted in the seminal *Capital Culture* and a broader, deeply rooted hyper-masculinity (Riach and Cutcher, 2014). However, it is important to remember that these cultures to not permeate in isolation but are woven into the very fabric of finance more broadly, as well as shape the spatial, material and technological frame in which they operate (Hirst and Schwabenland, 2018; Baeckström, 2022).

Affective relations of gender and space

The gendered construction of finance is, without doubt, woven within the contours of culture and practice that permeate through the City of London. However, it is equally important to view the material tapestry of architecture and space as both a container and reflection of gendered hierarchies, privileging a dominant and elite masculinity framed around 'virtues' of competition, aggression and risk seeking (Simpson, 2021). These global hubs of financial action are often indistinguishable across the globe, universally characterized by impersonal, glass and/or concrete buildings closely grouped together in a microsite, comprising spaces where there is little or no attachment to the geography around it. The buildings themselves radiate confidence and are configured to represent the social, economic and political wealth and power relations within. Together, market participants share a collective identity represented in part by informal norms as well as the physical space itself and where this spatiality is shaped by its occupants. As Smithsimon (2010: 2) argues, the exclusive design and formation of place 'encourages [its] residents to become dedicated defenders of the exclusivity' and, in doing so, 'enshrine, foster and reproduce exclusive attitudes among residents'. Similarly, Kaika (2010: 468) points out in relation to the City of London that while it is a public space it is one which is 'privatised, surveilled, surgically clean, overregulated and bare of public amenities', thus reflecting the needs of those who occupy the space as opposed to wider societal needs. Longlands (2020: 24) further developed this perspective, indicating that alongside

> The social institutions of the City (restaurants, bar, boutique hotels, private members clubs) … the space is still shaped primarily for and by the needs and everyday social practices of these highly paid workers

who occupy the office spaces. There are sandwich bars, a champagne bar, an optician's and a dry cleaner. There are gentleman's outfitters, a barber shop, a shoe shop, and a shoe repair shop.

In other words, the affective relations between individuals and the material form that surrounds them work together to produce an elective affinity that engenders a sense of belonging and, inversely, exclusion. The tapestry of wealth and exclusion is transformed and written onto the bodies, material surrounds and sense of meaning encoded into place to elicit a heightened sense of belonging as well as to define what is considered *in* and *out* of place.

The topography of dominance and masculinist encoding of both legitimacy and power that permeates through the City is designed to intimidate as well as excite, marking a boundary of, in Sack's (1993: 326) terms, *guest, stranger* and *citizen*. These symbolic layers written onto the urban topography are there for all to see and to feel as an emotive force of affect. As Spencer (2016: 140) reminds us, affect is that which 'circulates between one thing and another' to create a system of uncoded and prelinguistic expressions that resonate between the material organization or architectural form and the self. In other words, affect is founded on the emotive experiences and force relations that exist between and give meaning to our embodied and cognitive functions that shape movement, perception, thought and cultural practice as we move through the social and material world (Massumi, 2002; Gregg and Seigworth, 2010). Within the City of London, we can each move through its streets and encounter different frictions and responses to the surrounding material and social architecture, all of which can be captured within a process of bodily and emotive rhythms that constitute affect. This is to assert how urban design never communicates a single, static idea of culture or meaning, but produces a set of *affective relations* that connects individuals to a deliberately curated (if ad hoc) spatial environment (Moussavi, 2020). If we think of place as a version of 'text' containing and communicating meaning, then we can begin to examine the affective interpretations that blend and clash, forming the innumerable sense of presence and cultural expression as a multidimensional space of 'text'. This is an affective reading of space; one in which affect is transmitted by a form before being processed by individuals to produce affections of thought, feeling, emotion, moods and power relations, including gender.

Stepping inside financial institutions, we continue to see the interaction between gender regimes and place making. Gendered norms and spatial processes intertwine and underlie power hierarchies and homosociality, thus leading to a duality whereby investment management can be considered both a gendered and gendering space (Longlands 2020). For example, in most investment banks the fund management and trading floors (the 'pit') are hostile environments where aggressive communication styles and a macho, 'testosterone-fuelled' culture are normalized (Riach and Cutcher, 2014). Such spaces are governed by competitive practices where individualistic driven goals and a performance-led ethos embody the day-to-day norms and perpetuate male privilege and power structures. Despite the decline of the traditional open-outcry exchange over the last two decades and the increased technologization of finance, a masculine physicality, aggression and competitively driven 'battle hardened body' still remains prominent (Zaloom, 2006; Riach and Cutcher, 2014; Simpson, 2019). As one trader, who had been working in the City for 11 years at the time we spoke, encapsulated in an interview from a previous project, "I think I've worked with two women in my whole career, and one was a secretary" (Simpson, 2021: 59). This is, in other words a space designed for and by men of a specific class and race disposition (Baeckström, 2022), with the social practices in the City reinforcing a perspective that 'inequality is inevitable and a necessary consequence of meritocratic, "natural" market forces or the survival of the fittest' (Longlands, 2020: 50). The very drive for 'market perfection' is encoded with gendered understandings of competition, speed, aggression and, above all, survival.

In what follows, we draw on the voices of 12 respondents, each engaged in 'front office' finance work, that reveal the challenges, tensions and opportunities created by the changed spatial relations of their work. Of the respondents, all but one were men and were, at the time, engaged in working from home arrangements during one of the COVID-19 lockdowns. The participants were selected using a snowball sampling technique, with care taken to ensure participants were not known to each other. The participants varied in terms of age, ranging from late twenties to early fifties and in terms of levels of at-home care responsibility and were at different stages of their diverse careers: portfolio manager, senior analyst, dealer, chief executive. During the in-depth interviews, which took place online, each participant was asked to reflect on the move of work from the office to home as well as to reflect on how the physical space of work influenced their professional identity and how they did their job. Subsequently, participants were asked to explore the ways in which the 'doing' of finance work revealed itself as 'different' in the domestic sphere. The findings revealed that spatial processes were very different in the home versus professional environment. Gender relations and professional identities shifted significantly, with most participants now juggling home responsibilities within the working day. Overall, three key dimensions of discussion emerged from the interviews.

Reconfiguring the home

Spaces of finance present a very material allure, one which gets written onto the bodies and cognitive thought processes of finance workers. The City of London, for example, entwines a historical legitimacy, which stretches back to Elizabeth I, with the ever-expanding steel and glass edifices that dominate today's skyline. Bringing a coherent structure of oneness, wholeness and greatness, these vast steel-and-glass structures cast, in the language of Wacquant and Bourdieu (1993), a 'great reservoir of symbolic power' to underline the power and ubiquity of the market mechanism. The high towers of finance may create a narrative of boisterous, broad-shouldered masculinity, with their toughened, sealed and reflective glass veneers, but it is on ground level that the City lives and breathes. Come five o'clock, the throng of chatter changes to a more bullish and combative tone as the pubs fill up. Suited bodies pour out onto the street, their heads lost in the haze of cigarette smoke. Within this social arena, individuals coalesce face to face and, in doing so, give a very human and social face to the abstractions of finance. Communication technologies may have facilitated a degree of geographic dispersal of financial trading, yet there is still a need to access top technological, accounting, legal and economic forecasting expertise, located within centralized headquarters (Sassen, 2005; Longlands, 2020). The materiality and affective relationships that run through this space of work create a strong bond of affinity and are an integral aspect of professional identity. Yet, all of this was lost with the COVID-enforced restrictions, which led to discourses of loss and even mourning.

Recounting the physical spaces and the large, imposing buildings, participants reflected on being "kind of in awe" (Ethan, portfolio manager) when they enter the City of London and feel the authority symbolized by the hegemonic architecture. Others reflected on the influence of physical spaces on functional aspects of the business, such as developing client relationships. For example, Robert (investment analyst) spoke of how the "Mayfair bars and restaurants" were spaces where "private client deals and business took place [...] opposed to Canary Wharf which is more in your face". While the social geography of Mayfair and the City is characterized by 'old money' and historical legitimacy that comes with a sense of exclusivity, Canary Wharf contains an aesthetic of new, steel-and-reflective-glass buildings which are imposing, disconnected from the locale around them and closed off to public access. This, as Başeren (2010) argues, symbolizes the built form of neoliberalism, bringing into a material frame an architecture of prestige, exclusion and elitism as well as the global power brokers within. Here, the physical spaces of finance were seen as influencing self-image with notions of success, power and money aligned to the spatial and gendered frame of work. Spatial design and the urban landscape are both a product of and a process through which identity is formed, giving rise to a shared sense of belonging (Wacquant and Bourdieu, 1993). As such, moving away from these elite cultural places of finance, where "being a big swinging banker is part of the culture ... and it makes you feel great about yourself" (Noah, chief executive officer [CEO]), to the more mundane and domestic setting of the kitchen table or hastily constructed home office had a significant impact upon understandings of self.

For many, this meant trying to 'replicate' the original surroundings of the City within the home. As Henry (dealer) summarizes, "[I tried] to replicate it as much as possible. [...] It pretty much looks exactly the same as what it would be in the office", while Robert (investment analyst) notes that, "I just tried to replicate what I have at work so I felt like I was at work then". Much weight was given to the importance of technology and the lack of screens for their home set-up:

'Other things I missed [...], we have big screens that show where all our projects are. We have the wind speeds, how much electricity we're generating and things like that. It reminds you what you're doing at the end of the day. It reminds you sort of the assets that we're creating.' (James, CEO)

The link to technology as a 'window' through which the abstractions of finance become real is a central aspect of work. As Zaloom (2006) and Lépinay (2011) note, finance workers are 'plugged into' technological systems in a way that is emulated by few other sectors. Peering through these multiple screens, often stacked one on top of the other, and sitting side by side, finance workers extend their 'reach' to encompass a global finance market system of near instantaneous communication. It is how workers are 'connected', both to each other, forming a global marketplace, and to the object to which they are attached; in James's case, electricity price. Screens, therefore, were seen to be not only vital as part of the day-to-day continuation of finance work, but a very real window through which they could no longer peer to enact their occupational practice. For James, this came at the cost of work-based meaning. They were no longer connected to the impact of their labour and how their investment strategies could be translated into something measurable, such as wealth generation.

Wired into the market landscape in this way, technological systems are a vital tool for market communication. These systems map out, across multiple screens, rolling news that may impact on a particular market position, evolving data, market price and other metrics, helping to inform finance workers about key decision points and trades. Since the advent of electronic trading and the decline of the traditional open-outcry exchange, it is through such systems that finance works internalize, respond to and feel the rhythms of market movement (Zaloom, 2006; Simpson, 2021). Yet, the recreation of this environment at home often brought limitations:

'When you have a couple of screens, warning screens typically dedicated to, like, share prices and markets – then adjusting down into smaller laptop, you missed that information when you're doing something else. So, if there's a big market move and you're not on top of it, you miss it. It's an interesting difference between working at home and working in the office – having your screens.' (Robert, investment analyst)

Without the visual 'real estate', Robert could no longer 'feel' or 'touch upon' the market in the same way, limiting how a finance worker can weave between actual and digital space to conduct their practice through an unfolding system of technological adaption (Christophers, 2009; Mackenzie, 2018). If, as Simpson (2021) has argued, technology becomes a network through which the abstractions and ephemerality of finance can become real, measurable and calculable, without this, individuals such as Robert began to feel disconnected and, as a result, more exposed. The material and technological landscape creates a surface through which finance workers are able to 'feel' and be connected with one another, yet the atomization of working from home is more than simply relocating work. In other words, it is through the window of technology that finance functions and is a way of shutting workers off from their domestic surrounds. Removed from the chaotic whir of interaction within the City to the isolated space of the home, these finance workers struggled to recreate the same sense of connection or affinity. The affective reverberations of market 'noise' were not transformed into something more tangible, as happened in the space of work, leaving a greater sense of disconnection – from each other and the rhythm of the financial markets.

Managing the domestic

Beyond simply the technologies of work, the move to the domestic setting created the spatial challenges of setting up an office, commonly in a spare bedroom or sitting/dining room area. For others it was finding a space at the kitchen table which then had to be shifted during the day so other members of the household could avail themselves of it. This proved to be challenging and created a new frame of domesticity that had, hitherto, been separated from the field of finance as individuals faced competition for space and constant spillover into home life and compromise between family members. Afterall, finance is an industry that has been incredibly resistant to the working-from-home shift. While finance firms have invested heavily in ultra-low latency, literally reducing the material length of fibre-optic cabling between their offices and key servers to increase speed (Mackenzie, 2018), competition for space and internet bandwidth became new markers of domestic dispute – an issue especially pertinent for those with young children and/or elderly family members:

'If you have young kids in school or a toddler, trying to find some space in the corner of the kitchen or the sitting room to isolate – it's extreme difficult.' (Noah, CEO)

'It was difficult and feasible, just about. My wife works, I work, so we had two people working from home. I have three kids, one in school, and two in college. For a time, everybody was working remotely, or studying remotely, or attending college remotely. […] I'm in a dining room, the kids were in their bedrooms and my wife was in a box bedroom, which we turned into a home office. We were able to find space but just about.' (Ben, risk manager)

The trial of bringing together the space of work, school and home is far from unique to finance workers. However, it is important recognize the hitherto highly segregated spatial – and gendered – arrangements that function across the field of finance. It is an arena that went from the sharp use of techno-frontiers, from the visual real-estate mapping of the geometric world of financial interaction, from the race for ever faster, technologically accelerated communication systems that expand the possibilities of capital and speed of exchange (Mackenzie, 2018), to the more mundane challenges of 'finding a corner in the kitchen'. Stripped back are the techno-futurist possibilities of accelerated financial exchange, which privileges a masculinist vision of an impersonal and disembodied marketplace constructed around technological logic, and introduced is the general 'mess' and interruption of the domestic sphere.

As a consequence, many participants spoke of the need to assert boundaries to, once again, demarcate between the two spheres – the masculine environment of work and the routinized femininity of the home:

'I found I got more done when I went into the office for those two days 'cause there was peace and quiet in the office. At home my partner was working from home, my child was running around. […
But] in the office at work there was space and peace and quiet.' (Lucas, investment manager)

Here Lucas encapsulates the need for separation, dividing the world of work from home. Many offices in the City continued to remain open even during lockdowns, with some roles such as traders as designated 'key workers' and granted exemptions. This meant that some, such as Lucas, could continue to use the office for a fixed number of days a week. However, for those who were not granted an exemption, or for the remaining working days, the hitherto insulated and compartmentalized spheres of masculinist competition, aggression and risk seeking would now play out in the cramped and domestic sphere of the home. Stripped of the material trappings of importance and a long way from the expansive office environments of the City, our participants now faced the more domestic – and feminized – challenges of family life. For example, one senior market participant recollected how, "while waiting for the kettle to boil, I would do the dishes or stack the dishwasher" (Ben, risk manager). However, much of the discourse centred on the frustration of not being able to enjoy the 'perks' that the spatial frame of the City offers:

> 'In the London office [we have] two beer kegs and a champagne fridge and foosball and things like that. You do miss those types of things. After work, if you're having a really stressful day, you'll go pull yourself a pint and sit down, look out at the view. [...] So now, after the initial novelty of working from home wore off, [it became clear] that there is a reason why we created things called offices last century – to centralize people coming together working as a team and getting stuff done away from, say, the distractions of domestic issues.' (James, CEO)

For James, working from home means foregoing the elite markers of privilege – the view over the City or the champagne in the fridge – and now being confronted with the 'distractions' of domesticity. Whereas Hochschild (2012) outlined how, for men, the domestic sphere was a site of relaxation and a place to 'unwind' after work, requiring women to undertake the 'second shift' of facilitating this, we can see through James's quote how the office functioned as this environment. Surrounded by masculinist camaraderie, the office was the space to play foosball, 'pull yourself a pint' and unwind. Here, work is a space of seriousness and work, but also facilitates a gendered form of 'play' that cannot be obtained in the sphere of the home. In relocating to the domestic setting, lines of identity were now blurred between personal and professional senses of self, which in the financial context are not always aligned (Simpson, 2021).

The 'doing' of work at home

In moving the site of financial work from the City to the home, we can begin to ascertain the ways in which, as Longlands (2020: 67) argues, the spatial arrangements of finance help to reaffirm gendered understandings of identity and value that, in turn, help 'constitute and confirm particular forms of hegemony in global finance'. At the same time, however, the lack of a boundary or demarcation between the two fields created new demands to always be 'on' and available. This was frustrating for some who had care responsibilities and were unable to delineate between work and home. For example, James complained how "people [would] keep jamming things into your diary at all times", not allowing him any meaningful segregation between work and home (Al-Asfahani et al, 2022; Fan and Moen, 2022). In an industry characterized by long hours and late nights, there came a new expectation that workers would always be contactable and 'ready to go' if a situation arose:

> 'The definition of when you are off work is not as clear as it was [pre-COVID]. When I left my desk, I was off work. I had my telephone with me, and people could get in touch with me, and I was picking up e-mails. If you are working from home on your laptop, well, I've noticed staff tending to respond to queries at odd hours of the night; 10:00 pm, 11:00 pm, which indicates they're working very late or very long hours.' (Ben, risk manager)

> '[I]t's very easy to get up on a Saturday and do work, or Sunday, which I find myself doing a lot ... you feel you should keep working.' (Robert, investment analyst)

While work hours and culture within finance were long, there was at least a separation between the two fields of work and home. During COVID, however, these two became increasingly blurred. For finance, this comes with added challenges where there exist strong psychological tensions created of operating between the two distinct spheres of work and society more broadly, but home specifically (Zaloom, 2006; Simpson, 2021). It is an ontological tension (or separation) that, on occasions, drove finance workers to use pseudonyms at work to symbolically and psychologically split the subjective self from these two domains, minimizing what Bauman (2016: 110) calls 'the burdensome duties of managing communal bonds'. Now, however, the two are irrevocably entwining, limiting the possibilities of managing or creating boundaries of separation.

For Berardi (2009), this blurred distinction between home and work, self and worker, is prevalent through late-capitalist production. Whereas work was once a site of 'temporary death', where workers would 'come alive' when they returned to social spheres of home or play, increasingly our very sense of self, identity and value is entwined with what we do at work. Now separated from this field of work, and therefore identity, participants spoke of a loss of value and even self. For example, Henry (dealer) stated, "You forget that you're part of a larger organization in a larger market. You don't see anyone, all you have is a headset and a screen", while Francis (market risk analyst) noted, "I felt like, fuck it. I'm a chat bot. We are just very well-paid chat bots. Is this it?" Most starkly, Robert (investment analyst) characterized a typical day as "about 95 per cent business and then maybe, at the start or the end [of the day], it's like oh, the family – it felt almost forced". For these men, each of whom are fathers, there was a twinned alienation: an alienation, as Berardi (2009) describes, from the self that comes as a result of ever-increasing work hours and demands, leading to a formation of self-through-work, but also now an alienation from work itself. Untethered from the spatial surrounds of finance, an arena that, through its very construction, gives cultural legitimacy to the enactment of finance and, with it, gender relations, these participants felt dislocated from both *work* and *home*. Again, as Henry (dealer) describes:

'When I was in the office you worked your hours and that was it …, you walked out. But I found that when I was at home, you tended to stay around. We're sitting in the room and, "well, I don't really have too far to go, I'm trying just to get this done and I don't want to worry about tomorrow". [But] all of a sudden, it's an hour and half later and it's 7:00 pm and you're still sitting there.'

It is perhaps less that COVID created new obligations of work, but that without a spatial separation between work and home participants maintained their commitment to work longer and longer hours. This maintenance of 'always being on' underlines a continued dedication to the financial market mechanism. Where finance workers already channelled a total commitment to become at one with the speed and flow of the marketplace (Zaloom, 2006; Simpson, 2021), they now operated within a temporal boundary without end, that segregated themselves from being fully either 'at work' or 'at home'.

Conclusion

In this chapter we have examined the ways in which the physical workspace of work is deeply intertwined with understandings of self in investment banking and trading. The physical constructions of investment banks are symbolic, reflecting of and shaped by their inhabitants and the hierarchies and informal mechanisms and norms which govern the sector. We see here, as Spencer (2016) argues, that the production of a subjective self is, in part, an ongoing interrelationship with the spatial form, is integral to the creation of a collective identity and marker of belonging. That is to say, beyond aesthetics, the environmental apparatus of spatial and material affective relations of finance, specifically focusing on the material tapestry across the City of London, produces a distinctive set of mentalities and what Bourdieu (1977) calls dispositions that are conducive to the smooth operation of financial production. Bringing Bourdieu and affect together creates a prism through which individuals 'feel' and 'respond to' the history, culture and common assumptions that are co-constituted 'between skilled agent and pregnant world' (Wacquant, 2014: 5). Here, place is interwoven with affective histories that shape narratives of belonging, social hierarchy and cultural expression. It is here that we see our finance workers expressing an affinity with the placed surrounds of finance and how they struggled to adapt following the relocation to the feminine sphere for the home; which, in turn, carries its own affective relations. To pass through the slick and reinforced security that encloses financial architecture is to instil a sense of belonging, or citizenship, among its members. Established here is a symbolic, masculinist identity and unity between the material topography of finance and the suited bodies of finance workers. Yet, transported to the home for the duration of the COVID-enforced lockdowns, this symbolic relationship between the field, on the one hand, and the self, on the other, began to challenge markers of identity, belonging and even ability, while maintaining masculinist approaches to work, as seen through needs to separate the two spheres and their affective relations. After all, to be in the world is to feel the affective force of meaning, history and belonging, a sensory feeling that carries exclusionary and hierarchical gendered understandings (Massumi, 2002).

As the data reveals, during the pandemic the disconnect from these physical spaces to domestic realms led to a reflexive negotiation of self, often reflected in a dynamic multiplicity of roles – from banker to father, from portfolio manager to carer, from placing a £60 million trade to emptying the dishwasher while the kettle boils. Managing these multiple identities was challenging, with participants both consciously and unconsciously making sacrifices to one identity so as to engage with the other. Especially in the context of an industry that pivots around disconnected social relations and an insulated material and technological arena that privileges market 'values' of growth, profit and competition over social values of care, responsibility and reciprocity (Zaloom, 2006; Riach and Cutcher, 2014; Simpson, 2021). What we saw instead was a shifting set of affective relations that pivoted around greater emotional responses of doubt, uncertainty and, in turn, a continued commitment to be 'present' in the financial market sphere, despite the domestic setting. There was a need to be 'shut off' and demarcated from the domestic, establishing a detachment from the domestic space immediately around them and a continued commitment to virtual presence within the field of finance. Continuing to be 'wired into' the techno-social financial landscape (Lépinay, 2011; Berardi, 2012), these workers may have been physically located within the domestic, surrounded by the noise, chaos and mess that that entails, but they remained committed to gendered hierarchies and established privileges encoded within finance. Rather than remaining active as affective relations within the physical tapestry of the City of London, this was reproduced and replicated through virtual systems and (albeit limited) screens, enabling them to remain cognitively present in one field, while absent from (if located in) the domestic setting.

At the outset of the research, we posed the question whether the pandemic would provide an opportunity to challenge existing gendered hierarchies and spatial processes. The reality is that while the pandemic temporarily placed investment bankers and traders at the kitchen table, the power dynamics and organizational norms remained, albeit temporarily shifting into an online setting. The social aspects of work did disappear during the pandemic, due to the global lockdowns, but gendered hierarchies were not really challenged – they were simply muted for a time. Financial truths of subjectivity and its affective relations do not rule from outside or above, but are embedded within a common system of knowledge and practice that, as Spencer (2016) argues, become encoded into the spatial form. Equally, they are channelled through the systems of social interaction and technological connection that replicates the affective relationship between one another in a virtual, rather than material sense, and continue to reproduce masculinist practices of finance, even from the domestic sphere of the home.

References

Al-Asfahani, L., Hebson, G. and Bresnen, M. (2022) Reinforced or Disrupted Ideal Worker Norms in the Pandemic? Analyzing the Gendered Impact of the Pandemic on Professional Specialisms in a Professional Services Firm in Kuwait. *Gender, Work, and Organization*, 31(2): 644–665. DOI: 10.1111/gwao.12932.

Baeckström, Y. (2022) *Gender and Finance: Addressing Inequality in the Financial Services Industry*. Abingdon, Oxon: Routledge.

Başeren, Z. C. (2010) *A Baudrillardian Reading of Canary Wharf: Space as Economic Simulacrum*. Available at: https://www.academia.edu/7299925/A_Baudrillardian_Reading_of_Canary_Wharf_Space_as_Economic_Simulacrum.

Bauman, Z. (2016) *Strangers at Our Door.*. Cambridge: Polity Press.

Berardi, F. (2009) *The Soul at Work: From Alienation to Autonomy*. Los Angeles, CA: Semiotext(e).

Berardi, F. (2012) *The Uprising*. Los Angeles, CA: Semiotext(e).

Bourdieu, P. (1977) *Outline of a Theory of Practice*. Cambridge: Cambridge University Press.

Christophers, B. (2009) Complexity, Finance and Progress in Human Geography. *Progress in Human Geography*, 33(6): 807–824.

CityWire (2021) *Alpha Female Report 2021*. Available at: https://uk.citywire.com/Publications/WEB_Resources/alpha-female/alpha-female-2021-dollars.pdf (accessed 15 April 2022).

Collins, C., Landivar, L. C., Ruppanner, L. et al (2021) COVID-19 and the Gender Gap in Work Hours. *Gender, Work, and Organization*, 28(S1): 101–112.

Fan, W. and Moen, P. (2022) Working More, Less or the Same During COVID-19? A Mixed Method, Intersectional Analysis of Remote Workers. *Work and Occupations*, 49(2): 143–186.

Fisher, V. and Kinsey, S. (2014) Behind Closed Doors! Homosocial Desire and the Academic Boys Club. *Gender in Management*, 29(1): 44–64.

Gregg, M. and Seigworth, G. J. (2010) An Inventory of Shimmers. In: M. Gregg and G. J. Seigworth (eds) *The Affect Theory Reader*. Durham, NC: Duke University Press, pp 1–25.

Hertz, E. (1998) *The Trading Crowd: An Ethnography of the Shanghai Stock Market*. Cambridge: Cambridge University Press.

Hirst, A. and Schwabenland, C. (2018) Doing Gender in the 'New Office'. *Gender, Work and Organization*, 25(2): 159–176.

Ho, K. (2009) *Liquidated: An Ethnography of Wall Street*. Durham, NC: Duke University Press.

Hochschild, A. R. (2012) *The Managed Heart: Commercialization of Human Feeling*. Berkeley: University of California Press.

Kaika, M. (2010) Architecture and Crisis: Re-inventing the Icon, Re-imag(in)ing London and Re-branding the City. *Transactions*, 35(4): 453–474.

Lépinay, V. A. (2011) *Codes of Finance: Engineering Derivatives in a Global Bank*. Princeton, NJ: Princeton University Press.

Longlands, H. (2020) *Gender, Space and City Bankers*. London: Routledge.

Mackenzie, D. (2018) Material Signals: A Historical Sociology of High-Frequency Trading 1. *American Journal of Sociology*, 123(6): 1635–1683.

Massumi, B. (2002) *Parables for the Virtual: Movement, Affect, Sensation*. Durham, NC: Duke University Press.

McDowell, L. (1997) *Capital Culture: Gender at Work in the City*. Oxford: Blackwell Publishers.

McDowell, L. (2010) Capital Culture Revisited: Sex, Testosterone and the City. *International Journal of Urban and Regional Research*, 34(3): 652–658.

McDowell, L. (2011) Making a Drama Out of a Crisis: Representing Financial Failure, or a Tragedy in Five Acts. *Transactions of the Institute of British Geographers*, 36(2): 193–205.

Moussavi, F. (2020) *The Function of Form*. New York: Actar.

Riach, K. and Cutcher, L. (2014) Built to Last: Aging, Class and the Masculine Body in a UK Hedge Fund. *Work, Employment and Society*, 0(0): 1–17.

Sack, R. (1993) The Power of Place and Space. *Geographical Review*, 83(3): 326–329.

Sassen, S. (2005) The Embeddedness of Electronic Markets: The Case of Global Capital Markets. In: K. Cetina and A. Preda (eds) *The Sociology of Financial Markets*. Oxford: Oxford University Press, pp 17–37.

Sheerin, C. and Linehan, M. (2018) Gender Performativity and Hegemonic Masculinity in Investment Management. *Gender in Management*, 33(7): 561–576.

Simpson, A. (2018) Consecrating the Elite: Culturally Embedding the Financial Market in the City of London. In: T. Gleelan, M. G. Hernando and P. Walsh (eds) *From Financial Crisis to Social Change: Towards Alternative Horizons*. London: Palgrave Macmillan, pp 13–30.

Simpson, A. (2019) Establishing a Disciplining Financial Disposition in the City of London: Resilience, Speed and Intelligence. *Sociology*, 53(6): 1061–1076.

Simpson, A. (2021) *Harm Production and the Moral Dislocation of Finance in the City of London: An Ethnography*. Bingley: Emerald.

Smithsimon, G. (2010) Inside the Empire: Ethnography of a Global Citadel in New York. *Urban Studies*, 47(4): 699–724.

Spencer, D. (2016) *The Architecture of Neoliberalism: How Contemporary Architecture Became an Instrument of Control and Compliance*. London: Bloomsbury Publishing.

Tooze, A. (2021) *Shutdown: How COVID Shook the World's Economy*. New York: Viking.

Vyas, L. (2022) 'New Normal' at Work in a post-COVID World: Work–Life Balance and Labor Markets. *Policy and Society*, 41(1): 155–167.

Wacquant, L. (2014) Homines in Extremis: What Fighting Scholars Teach Us about Habitus. *Body and Society*, 20(2): 3–17.

Wacquant, L. and Bourdieu, P. (1993) From Ruling Class to Field Power: An Interview with Pierre Bourdieu on La noblesse d'Etat. *Theory, Culture and Society*, 10(3): 19–44.

Walby, S. (2015) *Crisis*. London: Polity.

Zaloom, C. (2006) *Out of the Pits: Traders and Technology from Chicago to London*. Chicago: University of Chicago Press.

5

What Is the Potential of Psychoanalysis to Understand the Relationship between Space, Objects and Subject Formation?

Darren T. Baker

Introduction

This chapter explores the potential of psychoanalysis for understanding the construction of identity through space and objects. More specifically, drawing on the two theories of 'transitional objects' and 'potential space' by Winnicott (1971), the chapter begins to explore the idea of how a subject's internal or psychic reality is shaped and supported by the objects and space around them. Objects and space blend psychic reality and fantasy together and can support subjects to take on new forms of identity in creative ways. There has been little research drawing on psychoanalysis and, moreover, Winnicott's ideas to understand how subjects reconceptualize, transition and move on from difficult or challenging moments with the support of objects, and the inherent tensions and contradictions in this process. By exploring transitional objects and potential space, scholars can begin to understand the close, 'micro' aspects of identity building as a way to position the role of material aspects of the organization in its construction. Psychoanalysis helps scholars to move beyond an understanding of space and objects as a set of individual and collective discourses, ideologies or meanings, and towards a deeper appreciation of how it shapes our affective life, our state of mind, who we are and, moreover, who we can become in the future.

The chapter offers a tentative psychoanalytic reinterpretation of Tyler and Cohen's (2010) influential paper on women and space in organizations. Using their study, the chapter shows the dialectic way in which identities and subjecthood are constructed through space and objects. For the women in their study suffering from marginalization and discrimination in the workplace, the negotiation of self with fantasies of the organization is a complex one that is mirrored and supported in their arrangements of objects and space. The psychoanalytic reinterpretation supplements the existing study but also opens up discussions about the importance of safe spaces for subjects, especially women at work, where they can make sense of the gendered challenges that they face, and how playing with objects and space can help to move forward identity construction and, moreover, create meaningful change in organizations.

Transitional objects

Winnicott was a British object-relations psychoanalyst who theorized transitional objects. Winnicott (1971) explains that transitional objects have an important psychic function as they support self-transformation. He developed his idea of transitional objects based on his work as a paediatrician and explained that infants develop an illusion that their parents are inseparable from them: parents psychically shield a child from the world by providing care whenever it is required. Winnicott (1971) describes how children develop a deep attachment with this early omnipotent view of the world, a world that is brought to them by their parents. The infant, in other words, cannot differentiate themselves from the mother or the external world. Ogden (1985), a psychoanalyst who has worked much with Winnicott's theories, explains that time, as a dynamic, ultimately disrupts this early attachment to the world; time detaches children from their parents because it provides structure and (societal) meaning that isolates a child further from its parents, and provides the child with a semblance of independence. This is traumatizing, however, because it involves the unconscious processes of splitting, whereby our sense of wholeness is divided by the realization that this psychic union with the parent no longer exists (or perhaps never really existed).

For Winnicott (1971), transitional objects for children, typically in the form of teddies and blankets, have a compensatory function that helps them overcome the maternal absence. A more inclusive understanding of transitional objects is that they help individuals to preserve, throughout the course of their lives, a sense of self through attachments with 'inanimate' objects. Transitional objects are not lost, repressed or mourned like so many things that we attach throughout life: they exist in a state of limbo between the unconscious, 'inside' realm and our shared external reality, enabling us to reattach ('cathect') to new objects in similar ways throughout our lives (Harrington et al, 2011). Transitional objects are therefore not part of our psyche, as they would be if they have been repressed, for instance, but exist in an in-between space that mediates our life. This specific ontological position of object-relations psychoanalysis is different from the Freudian focus on internal drives because it emphasizes emotional sustenance, and the interdependence between the internal and cultural worlds (Gomez, 1997; Woodward and Ellison, 2010). Examples of transitional objects include the humble television, a constant object in our lives that provides shelter while connecting us to the rest of the world (Silverstone, 1993), the mobile cell phone (Mowlabocus, 2016; Macrury and Yates, 2016) and the role of social media as a 'secondary' transitional object that plays into our unconscious need, or 'narcissism', for recognition and attention (Lasch, 1979; Turkle, 2011a, 2011b).

Bollas (1987) has elaborated on the idea of the transitional object as a site of self-transformation beyond childhood. According to Bollas (1987), it is the attachment itself with earlier, foundational transitional objects that is recalled in subsequent object-relating. In other words, transitional objects provide the 'affective' schema for attaching to objects, supporting motivation, hope and fantasy to the individual. One therefore does not seek a transitional object per se but surrenders themselves to it as 'a medium to alter the self' (Bollas, 1987: 14). Drawing on Krieger's (1976) work, Bollas understands *becoming* as an aesthetic process where the transitional object mesmerizes the individual, evoking the psycho-sensual 'feeling' of fitting *into* an object. Alexander (2008), drawing on Alberto Giacometti's sculpture of the *Standing Woman*, similarly argues that this is a form of immersion whereby the person becomes the object. Transitional objects therefore form part of the search for significance and meaning and reunifying with the prototypical mother as object.

Potential space

Potential space, for Winnicott (1971), extends the idea of transitional objects. It understands the immediate space surrounding an individual as an important one for play and symbolization, where external reality converges with internal psychic reality. Whereas liminal spaces constitute a complete retreat from reality, transitional spaces allow the individual to form new meanings, or re-entrench existing ones, that ultimately provide an opportunity for the individual to relinquish the present and transition into a new phase of life or state of being. When adults find themselves in a process of separation from an external object – whether it be a relationship, friendship, job or belief – it gives them an opportunity to ditch what Winnicott (1971) terms their 'false-self' and construct a new identity. The false-self is a term used to describe a defensive psychic structure, conformity, an evasion to pain or discomfort, and the desire to control and 'convince' others. For instance, when a relationship ends with another person, it evokes unconscious questions such as, 'Who am I?', 'What is my place or position here?' (for example, in an amorous connection or even in a team or organization), 'What contribution can I make?' Potential space is therefore a place where one can confront and distance themselves from conflict and antagonism and begin to digest and rework the self.

Through separation and disillusionment, Winnicott (1971) argued that 'creativity' and 'spontaneity' can emerge as part of a true-self where the individual can be playful, vulnerable and less contradictory. Separation can involve the realization that the individual cannot control the outside world and comes to understand it as both positive *and* negative, good *and* bad. Therefore, creativity emerges from a person's confrontation with reality rather than through omnipotence or fantasy (Dubouloy, 2004). In psychoanalysis, playing with fantasy gives the opportunity to accept reality and find agency. Jemstedt (2000), a Swedish psychoanalyst, explains that potential spaces emerge in part from the capacity of the mother to create a good 'holding environment' for the child, wherein she is able to attune herself to their emotional demands, what is sometimes called 'responsive mirroring'. A mother's gaze, in psychoanalysis, is important symbolically and psychically: it provides the child with the sense that they exist by reflecting their image back to them both literally in their eyes but also through empathetic gestures, movements and words. By containing a child's anxieties, the child is better able to make sense of their environment and feel safe, which enables symbolism to emerge from creative play.

Mirroring is required throughout life, not in a narcissistic sense but in the ongoing process of constructing one's identity. In potential space, one seeks to develop their talents for pleasure and satisfaction, as a way of discovering themselves. Locating transitional objects and an adequate holding environment can support an individual to transition more comfortably to a new phase of their lives. However, this process is not always a smooth one and can be fraught with anxiety and defensive regressions. Ogden (1985) builds on Winnicott's ideas of potential space as a 'dialectic process' and he describes four ways in which the capacity for potential space can collapse. First, *hallucination* is when omnipotence collapses and the person becomes trapped in a complete realm of fantasy objects. In this psychic position, there is no distance between symbols and the symbolized, and a person can struggle with forming ideas and feelings of their own. Second, when reality is used as a defence against fantasy it results in a foreclosure of imagination. For instance, in this psychic position, a boy is literally a boy and could never be a prince, even in fantasy. Therefore, there is little space for play and the possibility, or exploration, of meaning. The third case is when reality and fantasy are dissociated, and there is a foreclosure of the dialectic process because the person is concerned that this might generate feelings and emotions. In the fourth and most rare case, there is a complete foreclosure of both reality and fantasy. It is a form of coldness where experience does not register. Meaning is not foreclosed but is also not made.

Winnicott's theories of transitional objects and potential space are fundamental in the psychic development of a person beyond the childhood experience. Transitional objects resonate with our affective experience of unification with foundational objects, in particular the mother, during our early development; during adulthood, one seeks out this emotional experience through reattachment with other objects, especially during periods of self-transformation or change. Potential space is an unconscious holding environment for the individual *and* the transitional object and supports the individual to gain distance and protection from conflict, tensions and antagonisms in their wider life. Potential space is a place where they can start to recreate playfully aspects of themselves and their lives and begin reattaching again with wider objects, though this can be a fraught, tense and iterative process of collapse and the reopening-up between fantasy and reality.

Existing studies on transitional objects and potential space

There has surprisingly been little engagement with Winnicott's ideas in management and organization studies. Their application has largely been confined to the clinical and therapeutic context, but these studies are useful for management scholars to understand the application of these deceptively complex theories. BenEzer (2012), an Israeli psychoanalyst, explains how he developed a cross-cultural, creative therapeutic space on 'dreams' between himself, as 'analyst', and an Ethiopian patient. In Ethiopian culture, dreams are read as foretelling the future, but in psychoanalysis dreams are used as part of the therapeutic process to understand the present, or past, as an unrealized, unconscious wish or desire. For instance, when his patient shared dreams, BenEzer would ask him to associate with and predict possible future outcomes. This was therefore a mutually constructive space where the patient partly relinquished the more exacting tradition of dreams as foretelling the future with the support and clinical interpretation of the analyst through the psychoanalytic method of free association. Winnicott's ideas have also been used to a lesser extent in the humanities to understand our relationship with performative spaces and art. For example, drawing on recreations of city walks through *An Exeter Mis-Guide*, in the UK, Turner (2004) explains the guide as a creative invitation for walkers to imagine and reimagine what historical ruins in a future city could be made into; what real or fictional monuments could be erected; what meaning they would have; and to map their memories of places in Exeter and then contrast them with what they see when walking around. Turner (2004) argues how the city therefore becomes a playful place where the walk itself becomes a holding environment that people can use to trace and traverse boundaries of the city as real and fantasy.

Dubouloy (2004) uses Winnicottian ideas to explain how an MBA programme became a potential space for high-performing managers. Dubouloy (2004) suggests that there is a tendency for high-performing managers to become compliant, through processes of idealization, as they progress their careers up the ranks of an organization, which can reinforce a defensive 'false-self'. Dubouloy (2004) explains how, in order to recover their true-selves, managers need a reassuring environment to be more creative and innovative. Dubouloy (2004) shows how, during a two-year MBA degree, coaching sessions provided alongside teaching and learning took on the role of the 'good enough mother' for the students, and allowed fantasies and emotions to be worked through that in turn enabled the managers to deconstruct and reconstruct their sense of self in a way that prepared them better for leadership positions.

It is important for organization and management scholars to consider the relationship between worker identity and space through a psychoanalytic lens. Psychoanalysis offers a unique perspective on how identities can change and transform through object-relating and space. This is important in an organizational context because workplaces are deeply social and emotional spaces, where people unconsciously and consciously make connections with others, experience success and pride, but also endure difficult moments, including the experience of inequalities and oppression. This is even more important because many workplace environments could be considered the antithesis of many of the qualities of potential space. Potential spaces open up ambiguity, ambivalence, paradoxes and unconscious wishes that allow desires, fantasies and emotions to be expressed more freely. However, organizations struggle with this because it can empower voice, challenge and even resistance from employees. Organizations may therefore wish to suppress negative emotions and implicitly encourage the repression of emotions such as shame, loneliness and vulnerability among their employees through control systems, structures and cultural tropes (Taylor and Spicer, 2007).

Spaces that matter

This section presents a tentative psychoanalytic interpretation of the article 'Spaces that Matter: Gender Performativity and Organizational Space' (2010) by Tyler and Cohen. The article is a personal favourite of mine because the research data generated is emotionally laden, which lends itself well to a psychoanalytic interpretation. The intention of this section is not, however, to supplant the analysis by Tyler and Cohen but to present a supplementary analysis that can be used to show how psychoanalysis and, in particular, Winnicott's theories of transitional objects and potential space, can be employed by scholars.

Tyler and Cohen (2010) draw on the photographic installation *Grey Area* by Sofia Hulten to generate affective data in focus groups about the experiences of women at work. The installation is set within a stereotypical corporate office where the artist dresses in a grey suit and, across a number of photos, takes different positions and postures, including hiding and ultimately throwing herself away into a garbage-bin liner. The authors use this as a visual aid to begin discussion on the role of organizations, space and the challenges women face in constructing their identity at work. Psychoanalysts, particularly object–relation analysts, are known to employ artistic and creative approaches as both interpretative and therapeutic techniques. Winnicott (1971) himself drew on, for instance, a free-hand drawing technique called 'squiggles' with children during therapy as a point for discussion. Images can be an important way to help participants associate with emotional, unconscious aspects of themselves: they help to get around posing more restrictive questions, as part of a typical qualitative approach, which can evoke anxiety and defensiveness. Focus groups can also become a pseudo-therapeutic space for participants, who can learn from the experiences of others and help to build rapport and attachments.

Tyler and Cohen (2010) draw on Judith Butler's (2000) work on gender performativity to understand how gender is materialized through space in organizations. Butler's work is embedded in psychoanalytic thinking, particularly the work of Lacan, to explain how gender is achieved through performative acts that have the (unconscious) desire to project a coherent and believable gendered identity. In other words, it is necessary to recite particular gendered norms to be recognized as a viable subject: as either 'man' or 'woman' in the workplace.

Safety and ability to symbolize

In the focus group, one of the participants asked if she could take one of the pictures of the *Grey Area* installation back to the office:

'I really liked this last one [image], I mean I like it. I actually would like to take it away and pin it next to my computer or something. I know how stupid that sounds, but it's just like, it's a humorous kind of "Yeah, yeah, it's escape". It's stupid, but, you know, sometimes it's necessary to have that kind of reminder that ... to take the lighter side of it and just think, you know, you're asked [to do] too much sometimes. So I'm going to take it and pin it up.' (Tyler and Cohen, 2010: 185).

Tyler and Cohen (2010) explain that the vignette reveals how they had been able to set up an important reflexive space for the women participants, moreover, to raise consciousness of their gendered experiences at work. Winnicott would suggest that the research focus group became an important potential space for the women participants, wherein they could begin to make sense of and, moreover, start to symbolize, their embodied experiences as women at work. The focus group had likely become an important holding environment for the women participants: they felt comfortable enough to share similar experiences of unfairness and discrimination with one another. This is known in psychoanalysis as 'mirroring', when subjects are able to recognize themselves in others, which can generate a sense of safety. The women in Tyler and Cohen's (2010) study therefore felt safe enough to begin to make sense of the challenges and pain they had experienced in the workplace. In the potential space of the focus group, a photograph from the installation appears to have become a transitional object for one of the women, who asked if she was able to take it away with her to put on her own office desk. The photograph would likely help the participant in the future to make sense of herself, especially when faced with challenging or discriminatory situations. The installation, as a transitional object, therefore reminded some of the women of who they were – their true-selves in Winnicott's terms – and moved them unconsciously into a less anxious or defensive position where they could begin to conceptualize and make sense of their challenging, discriminatory workplaces.

Dialectical dissociation

Tyler and Cohen (2010) analyse how women experience the demand to be constantly available and hyper-accessible while rendered invisible at work:

> 'I think there's a demand that women especially are accessible. When you're talking about women being in offices where … where you can't hide away, I don't know, somehow I think as a woman you're expected to be always happy, happy, shiny, accessible for people to come and talk to, whereas I think men can get away with … telling you they're too busy to deal with your work.' (Tyler and Cohen, 2010: 186)

For one participant, Lisa, this sense of exposure was particularly acute because a male colleague would sometimes creep up behind her:

> 'I changed my desk layout a few months back … I'm tucked round a corner so my space is very limited … and I'd been sat with my back to anybody who was approaching my area and I realized I was very uncomfortable with this because whereas some people would speak

as they approached so I knew they were coming, there were a couple of … men who would come and stand and that was creepy … So, I rotated my desk round.' (Tyler and Cohen, 2010: 187)

Tyler and Cohen (2010) discuss how many of the participants identified with the photo of the woman in the installation throwing herself into the garbage. They explain that the participants saw this photo as an escape from the workplace where the woman in the installation was made to feel worthless, a failure and almost inhuman in the masculine corporate environment where she worked. Reading the vignettes, Ogden (1985) may suggest that for many of the women in Tyler and Cohen's (2010) study there are different dialectical processes in operation between their internal selves and organizational space. For instance, he may suggest that there is a dissociation for some of the participants between their individual experience – what is sometimes referred to as 'psychic reality' – and the organization as 'fantasy', as a set of collective and individual constructs, ideologies, symbols and meanings of the workplace. The implication of this is that women feel individualized *within* the fantasy. In other words, the women feel that their experiences are somehow unique and distinct from, for instance, the fantasy of the organization as 'meritocratic' and fair. This would perhaps point towards the identification by the women with the photo as feeling worthless and invisible: in other words, it is not discrimination that has created their feelings of worthlessness but themselves – the women see themselves as not good enough, as 'lacking' in the fantasy. Winnicott would likely point to how Lisa had to develop (understandably) a highly defensive structure in response to the 'creepy colleague'. What is concerning about Lisa's circumstance is that the collapse of fantasy is challenging because it does not allow for internal digestion of what is happening in the outside world; the implication is that Lisa is unable to recognize fully and find the energy to challenge these behaviours as forms of harassment. Instead, she is simply left to defend herself by positioning her desk and desk objects to identify when he is approaching and prepare herself.

Potential spaces

Tyler and Cohen (2010) explore how many of the women curated their desks according to the implicit rules of the organizations:

Carole: 'I mean it sounds a bit calculating, but it depends. If I think it serves my purpose between not to show them [she keeps photographs in her desk drawer] and if I think at the end of the day I'm more likely to achieve the aims I want by not revealing them, and the depth of my emotion, then I'll do that. Sometimes though it really gets to me and they come out. [What gets to you?] The control, it just feels like leading a double life.' (Tyler and Cohen, 2010: 188)

Ogden may suggest that Carole is experiencing a form of 'hallucination', where her sense of self becomes almost completely trapped in fantasy. This is not a creative, individual sort of fantasy but the collective fantasy of the organization and, more especially, its explicit and implicit rules that govern how employees, and women, are expected to behave. The fantasy has taken hold of Carole to the extent that she has to micromanage the arrangement of her desk in order for it to align with fantasies of the organization as 'professional': ideas of professionalism have a masculine, gendered subtext where employees must not express emotion or reveal aspects of themselves. This psychic position means that Carole struggles to form clear ideas and feelings about herself in relation to the rest of the organization. In other words, the fantasy of the organization is all there is: there is little or no space for her to play creatively with aspects of her identity and the fantasy of the organization in ways that not only conform, subvert but produce new forms of identity and pleasure for Carole.

Many of the women, unsurprisingly, appear to struggle to use space in a way that would enable them to express themselves emotionally. However, Tyler and Cohen do use a few vignettes where women were able to shape desks and desk objects around them without inhibition:

> Susan: 'I think it was just to have my kids near me I suppose. I guess it shows people I'm human and that I'm approachable. That's what I'd like them to think. You know, it's not ... It wasn't a conscious effort on my part to make me seem mumsy or anything, you know. It's funny you should ask me about it because I'm noticing all these things now – like there's a butterfly I made out of my son's hands [using handprints] over there as well – and just like work is my escape and yet, I've surrounded my office with all these things that remind me of them.' (Tyler and Cohen, 2010: 188)

Susan shows how displaying family photos reminded her of who she was outside the organization and helped her through the (unconscious) sense that she did not belong in her workplace. Susan is not using reality as a defence against fantasy, as there appears to be an attempt to imagine and reimagine who she is as both 'worker' and 'mum'. In psychoanalytic terms, whereas there appears to be anxious splitting in the previous vignette – Carole defensively positions herself within fantasy as Carole 'the worker' – Susan appears to play with both the reality of her workplace as 'escape' (a space for her to be independent, presumably) as well as her role as mum. During the vignette she suggests the unintentional use of her son's drawing as a transitional object in her negotiation with the expectations of the workplace. In other words, she is able to hold onto both aspects of herself, and that she cannot control either world, and finds a degree of agency and comfort between both through the use of her desk objects.

Conclusion

The tentative psychoanalytic reinterpretation of Tyler and Cohen's (2010) influential article on women and space in organizations offers additional insights into the relationship between their internal selves and how they are shaped reciprocally by the objects and space around them. This moves beyond an analysis of the meanings and discourses that are attached by individuals and groups onto space and objects, to a deeper understanding of how physical objects begin to shape not only how we view the environment around us but also how we feel and, moreover, who we are. The study by Tyler and Cohen (2010) points towards a more tense, dialectic process in Ogden's terms, between fantasy and reality: the negotiation with self and organization is a complex process and this can be frustrated, as the study shows, when women face challenges of unfairness, harassment and discrimination in the workplace. The study reveals that women understandably become defensive when they experience inequality in the workplace, but that space and objects play an important role in supporting women to protect themselves when they face difficult situations or individuals. In other situations, the women were more successfully able to employ inanimate transitional objects as a way to shape a potential space, make more sense of the organization and their experiences and begin to shape new workplace identities.

The chapter has highlighted how Winnicott's (1971) theories offer scholars a way to connect space with our internal lives, beyond just an appreciation of discourse. Understanding how culture (organizational cultures) is generated through the individual, at a micro 'psychic' level, is important in understanding how cultures can be buttressed, lived through but also subverted, and made sense of. Transitional objects open up new discussions and possibilities about the role of space and objects in workplace identities and, moreover, in discussions of women's equality. Workplaces ought to be spaces where women, and others, do not encounter hostility, where they can recreate themselves in ways that are not only healthier but also enable more effective work. Women need a safe space to make sense of the challenges they have faced and to work through them both as an individual but also collectively, as a way of moving forward. This opens up ideas of play, playing and playfulness in the workplace, what sort of feelings are evoked, how we communicate ourselves and how we can work with and through space to create equitable change.

References

Alexander, J. C. (2008) Iconic Experience in Art and Life: Surface/Depth Beginning with Giacometti's Standing Woman. *Theory, Culture and Society*, 25(5): 1–19.

BenEzer, G. (2012) From Winnicott's Potential Space to Mutual Creative Space: a Principle for Intercultural Psychotherapy. *Transcult Psychiatry*, 49(2): 323–39.

Bollas, C. (1987) *The Shadow of the Object: Psychoanalysis of the Unthought Known*. London: Free Association Books.

Butler, J. (2000) *Gender Trouble*. London: Routledge.

Dubouloy, M. (2004) The Transitional Space and Self-Recovery: A Psychoanalytical Approach to High-Potential Managers' Training. *Human Relations*, 57(4): 467–496.

Gomez, L. (1997) *An Introduction to Object Relations*. London: Free Association Books.

Harrington, C. L., Bielby, D. D. and Bardo, A. R. (2011) Life Course Transitions and the Future of Fandom. *International Journal of Cultural Studies*, 14(6): 567–590.

Jemstedt, A. (2000) Potential Space. The Place of Encounter between Inner and Outer Reality. *International Forum Psychoanalysis*, 9: 124–131.

Krieger, M. (1976) *Theory of Criticism: A Tradition and its System*. Baltimore, MD: John Hopkins University Press.

Lasch, C. (1979) *The Culture of Narcissism: American Life in an Age of Diminishing Expectations*. New York: Norton.

Macrury, I. and Yates, C. (2016) Framing the Mobile Phone: The psychopathologies of an Everyday Object. *Communication and Media*, 11: 41–70.

Mowlabocus, S. (2016) The 'Mastery' of the Swipe: Smartphones, Transitional Objects and Interstitial Time. *First Monday*, 21(10).

Ogden, T. H. (1985) On Potential Space. *International Journal of Psychoanalysis*, 66(2): 129–41.

Silverstone, R. (1993) Television, Ontological Security and the Transitional Object. *Media, Culture and Society*, 15(4): 573–598.

Taylor, S. and Spicer, A. (2007) Time for Space: A Narrative Review of Research on Organizational Spaces. *International Journal of Management Reviews*, 9: 325–346.

Turner, C. (2004) Palimpsest or Potential Space? Finding a Vocabulary for Site-Specific Performance. *New Theatre Quarterly*, 20: 373–390.

Turkle, S. (2011a) *Alone Together: Why We Expect More from Technology and Less From Each Other*. New York: Basic Books.

Turkle, S. (2011b) *Evocative Objects: Things We Think With*. Cambridge, MA: MIT Press.

Tyler, M. and Cohen, L. (2010) Spaces that Matter: Gender Performativity and Organizational Space. *Organization Studies*, 31(2): 175–198.

Winnicott, D. W. (2003[1971]) *Play and Reality*. London: Taylor and Francis Ltd.

Woodward, I. and Ellison, D. (2010) Aesthetic Experience, Transitional Objects and the Third Space: The Fusion of Audience and Aesthetic Objects in the Performing Arts. *Thesis Eleven*, 103(1): 45–53.

6

Affecting a Desiring 'Woman Worker': A Spatial Interpretive Ethnography of a Café in India

Rajeshwari Chennangodu and George Kandathil

Introduction

In this chapter, we explore the process of producing an urban café in India as a 'women-only' workspace which claims to empower women through employment and entrepreneurial skilling. Our interpretive ethnographic exploration comprising participant observation of the kitchen space and the dining hall of the café reflexively analyses various notions of 'women' and 'worker' and the interconnecting affect that acts upon these notions and is acted upon by them within the production and embodiment of space. Drawing on Deleuze and Guattari (1988), our spatial analysis of affect not only exposes the dominating neoliberal forces that prioritize logic of commodification over alternate logics but also shows how these forces invest human bodies with an intense desire to become a 'better woman worker'. This occurs through the acquisition of skills that are submitted to the hierarchized logics running through various bodies. Thus, we argue the following: (1) a 'woman worker' performs a desire to connect with other bodies that are constructed as higher and better than other bodies, the predominant metric used in creating this hierarchization being proximity of the bodies to the market and its embodiment of market logics, and (2) a 'woman worker' who attempts to move closer to the market logics and embody them as neoliberal forces tends to become dominant in the café space. However, within the kitchen space, there are also sporadic intense desires to connect with the bodies who are engaging in solidarity movements, connections and embodiment through sharing joy, precarity and support which militate against embodiment of market logics.

Foodwork is often seen as a way through which women express and connect with social organizations such as families (Brenton, 2017). As reorganizing of care and productive work is emerging in urban areas, foodwork is also shifting rapidly into a form of paid work, carried out in gendered eateries where men occupy the positions of chefs and cooking staff (Ray, 2016). Against this context, we explore a café in Western India which claims to be 'women only', where women engage with the emerging space of commodified foodwork by selling their foodwork-related skills and cooking. During our ethnographic engagement we worked with the café workers, whom we found mostly essentialized as women, yet also create a multitude of feminine subject positions through their different desires of engaging with others as workers, entrepreneurs and food makers. Specifically, we explore their spatial construction of desirous work (Deleuze and Guattari, 1988), where the women engage and create a space of and for their work through creating multiple affective positionalities, which weave together as a café space.

Desiring foodwork

Desire, for Deleuze and Guattari (1988), is not a psychological state that people possess. Rather, it is a force that scans the past from the perspective of the present in search of possible combinations to actualize. In this sense, desire is a 'machine in the becoming' through the creation as well as breaking of flow of energies in the way it connects with other machines or entities. Thus, desire does not emerge due to lack of other machines or objects, but in connecting to other machines, and it is inevitably expressed in institutions and codes which are contingent and hence susceptible to change. Anyone working in their domestic kitchen may embody the desiring machine in multiple ways; for example, through desiring to be a 'woman' (a coded female human body) in a specific way, or a foodworker (a re-coded human body). The desire lies not in seeking to be a foodworker due to any feeling of lacking something but in the potentiality of the individual in embodying the connections between, in this context, her female body and the familial role of foodwork, and then with the various other objects, plans and intentions like skills, cooking utensils of various kinds and ingredients. Thus, desire works in the flow of energy that connects these various parts of the desiring machine, then proceeds in connecting with other machines like the family machine. Desire is not the one-time interaction that the female foodworker makes with the kitchen and other family members, but it is in the possible multiplicity and repetitions of such connections, which need further flows of energies, like in learning new skills to cook, in finding better places to buy ingredients and tasting new dishes so that she can cook them herself. Desire creates such potential connections for the 'bodies without organs', or the bodies that are not organized, or connected in a particular way so that there is a particular territorialized organization in the becoming (Pedersen and Kristensen, 2019).

For Deleuze and Guattari (1988), the desiring machine channelizes the flow of desire-energy in certain directions, creating a possibility of constituting a specific production machine, such as a café space. The channelization of flows of energy changes the desiring machine when they are connected to other social machines of production, creating new configurations. But, in return, the desiring machine stabilizes the flows of energies along with the other unconnected machines in the body without organs or the unorganized or yet-to-be organized machines, constituting a temporal organization from the multiple possibilities of connecting the various machines. Since energy flows continually, its channelization and connection involve the breaking of certain existing connections. In other words, stabilization and destabilization occur simultaneously. The temporal organization, simultaneously stabilized and destabilized, operates as a machinic assemblage of connections where various other connected machines emerge as a temporal machine. Stabilizing the connections of various machines into an emerging machine creates specific configurations and thus coded contingent boundaries, which for Deleuze and Guattari is part of territorialization. As stabilizing and destabilizing are simultaneous, so are the attendant territorializing and de-territorializing. By implication, the emerging machinic assemblages are not entities that exist, but are processes that keep changing. Since the connections that territorialize, for example, the female body as a machine of production of food are temporal and contingent, as these machines are stabilized, or territorialized, they can also be de-territorialized. The same female body can be connected to other machines of social production, through other desire machines, like that of a machine of a female corporate worker who visits a café where a female food worker works. Thus, desire becomes the fundamental system of connecting and disconnecting various other machines of social production, where flows of energies are channelized, and thus constitute not only concrete territorializations but also meanings and abstractions that emerge out of these territorializations, which are inseparable from production machines. De/ re-territorialization connects desire to the political economy of the market by designating the investment of energy in specific areas of the body and the economy, the withdrawal of such investments and their reinvestment elsewhere. These investment processes are guided by coding and de/recoding.

These territorialized assemblages or configured connections create affect. Affect is not an individualized psychological state but a material, embodied becoming, where the bodies are capable of being affected and of affecting other bodies (Ringrose and Coleman, 2013). Since affect is capable of changing the bodies, it can create new connections. Academic conversations that created an 'affective turn' (Beyes and Steyaert, 2021) have detailed the various shades of affect in organizational life. For example, the varieties of ways in which affect works include love and its absence creating gendered organizational experiences (Acker, 1990; Vachhani, 2015); affective solidarity (Hemmings, 2012) creating ethical experiences in the context of feminist resistance (Vachhani and Pullen, 2019); and affective generosity and resistance creating organizational ethics as an ongoing project emerging out of embodied connections of everyday organizational lives. Such studies mobilize affect as an active, agentic organizational participant that 'defines relations among all kinds of bodies' (Beyes and De Cock, 2017). Thus, affect becomes an integral and active part of relational organizing. It conditions organizational relations as it is embodied in organizational life in the 'in between' (Beyes and De Cock, 2017). Thus, affect is seen as a conditioning force in organizational life. In our empirical site, we observe affect through the embodiment of a café space, as a force that shapes the organizational experiences, where desiring bodies as both affected and affecting, become part of the hierarchies that emerge while placing themselves in the process of territorializing themselves into a stabilized space called café and continuously (re)organize or (re)territorialize their dreams, aspirations and resistance, as explored in the following sections. Further, Deleuze and Guattari's toolkit of machinic enslavement suggests that the contemporary capitalist system, which generates neoliberal modulations, produces and configures human subjectivity in two ways: (1) political forces of social subjection construct individuated and corporeal subjects, imposing particular values and identities onto persons within the system, and (2) the processes of machinic enslavement deterritorialize the same persons, robbing them of their identifiers and turning them into mechanic components of productive assemblages. Developing these insights further, Lazzarato (2014) considers pre-emptive production of subjectivity and machinic enslavement as central to neoliberal capitalism.

Methodology

The project that this chapter draws upon and engages in interpretive spatial ethnographic exploration of Café Kasturi (CK), located in an urbanized city in a western (federal) state in India. The women who worked in the café were from different socioeconomic classes and castes. Many of them had migrated from other (federal) states and rural parts of the same state, mostly with their male partners who relocated in search of better job opportunities, or these women were married to families settled in the city. The café was run by a self-help group which identified itself as a trade union for empowering women for livelihood gain. In this spirit, the café was being performed as a dedicated space constructed for women only, as a safe space for training and creating job opportunities for women.

As part of participant observation for her doctoral thesis, the first author worked (without payment) in the café kitchen from January to March 2019, visiting the place on alternate days, and was involved in some daily tasks like serving and helping the chefs and cooks in the kitchen, and shared with the billers her knowledge of performing simple tasks using the computer. The interactions with the kitchen workers were spread across morning and lunch hours, creating familiarity and friendships with the workers, which helped in casually conversing about their work, families and other spheres of life. They often shared their aspirational desires comfortably in their workspace in the presence of the author. The first author expressed her interest in engaging with them in a way that would make sense to all the participants, including herself. This needed a different kind of framing of her work and her experiences. Hence, she explicitly mentioned that her intention was not to earn and become a professional foodworker, but to write about the work they did as part of a book that she might write in future, as it was not easy to collectively imagine why and how a doctoral student, who had better potential to earn, was working in a café, unpaid. Such attempts to explain her positions to the workers created a space to talk about their ways of spatial embodiment of their desire by constructing the café through working and socializing in the café, becoming café workers, entrepreneurs and colleagues.

Stories of new desires

KC emerged from the foodwork that female foodworkers performed both in the café and in their family. They engaged with the family machine, becoming care workers (Aronson and Neysmith, 1996), while in the café they became employees, embodying a desire to engage with the café machine (Deleuze and Guattari, 2009). The café machine emerged through (re)territorialization of various machines that came to it through these women. For example, cooking and cleaning, which had already been territorialized as the family machine, were re-territorialized as the café machine. Thus, as the women workers embodied the skills to cook, clean and serve, they became cooking, cleaning and serving machines, which were then channelized into a café machine through the workers' desire to engage with it. In other words, the desiring machine emerged in the connections that were formed as these connections energized workers' skills of cooking and connected them with food as makers of food and foodspace. Thus, the desiring machine engaged their bodies in foodwork and imagined the possibility of creating new engagements and connections that would energize new skills such as entrepreneurial skills.

'I have a degree in Sanskrit, and wanted to continue studying, but then I was married off and came to this city. Now I am working, I have to be independent, my daughter will soon go abroad. I might join her in the future, run my own Indian café there using these skills I learn here to manage a restaurant. Indian restaurants will have demand abroad, so, this place and the experiential learning it offers matter.'

Vijaya, a supervisor at the café, explained her desire to engage with the café, relating it to her career and familial aspirations, a strong desire that energizes a specific goal. Her desire to engage with the café was channelized through such aspirations and in turn created for her an aspirational affect. Thus, she desired to engage with the café by becoming (or spatially territorialized as) a supervisor. Her engagement with the café began when the parent self-help group (SHG) decided to utilize the space, which was donated to them by a prominent scholar and social activist who worked for women's empowerment. The parent group thus desired to create a café that serves 'healthy food produced by women workers', on one the hand, and, on the other, would create a space that was run through 'the foodwork of women only empowering them', essentializing the process of constructing the workers' bodies as 'women'. Thus 'womenizing' the worker bodies was being essentialized in the claims of the café.

Accordingly, in the restaurant, cooking and cleaning skills that are usually assigned to and embodied by women were shaped and utilized in such a way that they could earn. One of the desiring machines connected women who were from underprivileged backgrounds with the women who desired to empower them. Within this relational connect, both *underprivileged* and *empowered* became affective states that shaped the experiences of the 'womanized bodies' that desire simultaneously their growth and the café's growth, coupling both: "I used to stay till 10 pm in the night, when this place was not this safe, it was covered with bushes, there was no building as big as this. I used to feel afraid, but I worked hard so that this place grows into an established café."

When their bodies and their work were being (re)territorialized into a café, that was an addition to managing their familial roles, which required them to perform care work, essentializing them as primary and unpaid care workers of the family (Bhattacharya, 2017): "We bring food from home, I have to finish cooking for my daughter before she leaves for her college every day. I finish cooking breakfast and lunch and pack lunch for myself as well." "Usually, my daughter-in-law cooks food, but she has gone to her parents' place because of her pregnancy, so I have to cook for the family."

These workers continuously saw themselves as women in all these conversations and were unquestioningly accepting of their roles as foodworkers and caregivers in the café as well as for their families. The roles they played in the café were in continuation with the familial caregivers' roles. Thus, the workplace that they wanted to create had to accommodate their caregiver role, besides helping them to earn for the family using their skills: "My children are in my home state. If I also start earning along with my family, we can educate them better by sending them to better schools."

Experiences of training channelized the desiring machines towards engagement with market dynamics, creating market interest and responding to these interests by becoming financially independent and entrepreneurial. Often, these ideas were shaped by the café managers and the parent SHG, which had their own plans to "transform the underprivileged into entrepreneurs":

> 'They can learn to use their existing cooking skills to earn money. This will create financial and job security for them. In addition, our values include providing food security by using traditional skills of cooking healthy food and offering such food to consumers. The women get trained in that aspect as well.'

In this way, the roles that the women played outside the workspace were being essentialized now through the café space. While the women workers wished to work and create a space for their work and earnings, the manager and the parent organization claimed that they want to "uplift more women from rural and poor backgrounds to entrepreneurship". These intentions converged to create a café space, where claims of empowerment intersect the flow of desire to engage with market opportunities to earn and overcome poverty, creating and connecting various feminine subjectivities while centralizing market logic as the metric of valuation. The managers and workers were coded and crystallized (essentialized) as women in this workspace, by each other and themselves. The workers aspired to be managers through the emerging valuation system. These aspirations were affective, as they shaped the ways in which managers and workers valued themselves and connected with each other, thus shaping the desire for engaging with each other. The affective aspirations further created certain positionalities – the ones closer to the market – as better valued and higher. Next, we explore these desirous embodiments of female subjectivities emerging into hierarchized positionalities that constructed the day-to-day café space through their movements, occupancy and spatial arrangements.

(Re)territorialization of space: the emergence of foodwork and workers

The café became a territorialized space, as various decoded and re-coded subject positions emerged through the interactions among its participants. Through their work, they engaged with each other and other material spaces, creating a space of cooking, serving and eating food – de- and re-coded as a space of foodwork – which became one of their many ways of spatial engagement.

> 'We make traditional thalis,[1] we make and serve them in healthy ways. We knew how to cook these items before coming here, because roti and dal is part of our home food as well. When we cook, we all become part of cooking and serving processes. We are trained to make biscuits also, many people come here for training as well.'

We see here a resurfacing of the 'empowering discourse' (Hickel, 2014) of healthy traditional food for the consumers – a claim that we heard earlier from the parent SHG and the café management. Thus, there emerged a collective assemblage of empowerment that connected the machinic assemblages of family and work and which involved numerous de/re-codings and associated de/re-territorialization. In short, the space became a café through various interpretations of the participants, who became *workers* creating and occupying the de/re-coded subject positions of empowerers, empowered, poor, modern, cooks and chefs and cleaners, and so on, when they engaged in interactions that became relevant to themselves and to the de/re-territorialized café. Each object, human body and the space that was created through placing the object or movements of human bodies became part of a territorialization process, where the flow of desire to engage in specific ways (re)signified the objects and human bodies in relation to foodwork. The space where tables and chairs were arranged for the consumers became the space where the foodworkers performed their hospitality and serving skills for the consumers. They performed other parts of the foodwork in the kitchen. The wall between the dining hall and the kitchen became a boundary that separated the performances of kitchen work and the process of consuming foodwork and food, by restricting consumers from accessing kitchen space.

As the de/re-coding stabilized and the investment of desire-energy repeated, the participants become workers, they embodied the foodwork, (re)created roles and territorialized space for each role through their body movements and interactions with the others. For example, the chef, Amala, stands between the tables used for vegetable chopping and the industrial stove. This space helps her to move between the cooking process unfolding on the stove and the preparation of ingredients occurring on the table next to the stove. In turn, her movements and interactions with the table and other material bodies created a position of the chef, along with her expertise to energize the position by engaging with more bodies in ways that are signified as that of a chef. The space where she moved became the space of the chef, where chef, a socially invested position, is created through embodiment as a spatio-material positionality. Further, *she* becomes a chef standing in this position and moving around it, mutually constructing her embodied self and the space as part of a repeated interaction, structured as chef. Similarly, the cleaning workers, who performed cleaning of used utensils, tables and ground, created and crystallized their roles through repeated investment of the desire-energy and embodying movements, engagement with objects like used plates, brooms and the human bodies that use the space and objects in foodwork and consumption of foodwork.

As these objects and bodies were territorialized into a food workspace, and the workers negotiated around this space and moved from their territorialized spaces into other such spaces, the desire-energy was reinvested in a specific area of the economy: the machinic assemblage of the market and its assemblage of collective enunciation. For example, better pay signified better value or skill (a specific market logic) in the café. Sangeetha, one of the two cooking helpers in the café, argued for better payment for her new work (compared to the old work of helping), which is vegetable chopping and occasional cooking, signifying payment with respect wherein better payment should translate into more respect for the earning body (another market logic): "I want to get the deserving amount [for my new work]. I am here to earn money. I am working diligently, performing all my tasks. I need to be respected."

The organization of the space emerges from these negotiations and affective intensities, often related to an embodied market logic, which create movements and spatial intentionality to move. Such movements and intentionalities constitute these spaces differently for different positionalities. These intentionalities conflicted with each other, materializing as arguments. For example, while most of the women refused to stay beyond 8 pm to cook and serve, the managers had to find someone who was willing, or force someone to stay back. Some women demanded extra payment for staying back, while managers refused to oblige, rationalizing that it was a work that workers need to offer for the growth of the café, as a part of the job. Such conflicts often led to workers leaving the job, since staying late into evening hours clashed with their familial care work hours. Thus, the marketized logics, which were encapsulated in the affective intensities of aspirations, affected and were repeatedly invested in the working bodies de/re-coded as managers and workers. Specifically, these logics affected them to engage aspiringly, and more with bodies which were closer to market, channelizing the desire to move closer to the market and hierarchizing the positionalities according to the proximity to the market. This was signified by the value and appreciation accorded for opinions, skills and suggestions of working bodies.

'Maya [a manager] is very strong. At the same time, once in a while, she even cooks. She comes and teaches us to cook South Indian food. She does not shout at us, there is no need for that, people obey her and respect her, she has that kind of influence on them because she has been managing the café, dealing with the consumers and the self-help group and training them.' (Vijaya about Maya)

'Everyone cannot talk to the consumers, we have to know their language, some of us cannot speak that welcomingly, or do not know English, there is a different way to speak to the consumers, the managers and supervisors can do it.' (Amala)

'You see, they are yet to learn many things, like talking to the consumer, serving the way they want, and understanding their needs. These women are not from that kind of background, that is why we need such training to make them better gradually. They can only cook. That too they cook the food that their families eat. They have to learn to be better. Some of them have improved quite a bit after coming here.' (Maya)

Similarly, kitchen workers felt that they should try to improve their skills and the production process for generating better income opportunities, and "better valued" skills: "I want to learn to deal with computers. Also, now I converse with the consumers, engage in small talk. I can respond to their queries comfortably ... Where else do I get such opportunities (for growth)?"

Drawing on Deleuze and Guattari, the stratification of bodies along with their embodiments and connections of positionalities were emerging through the interactions of the two types of machines, the concrete machinic assemblages and the collective assemblages of enunciation (for example, the discourse that centralizes and prioritizes market logic), creating hierarchies. The closer a body and its embodied skill to the market, the more it was considered confident, strong and improved:

'I used to cook, I have done a bachelor's and master's degree too. Here I built this café from scratch. I started with cooking, moved to supervision, and now go to client locations to serve tiffin. I talk to consumers of all kinds comfortably now, and that skill helps. That is how I become better; gain confidence. Similarly, many women come here to learn cooking in new ways, they can make roti and dal, but they learn to make things that they can sell in the market. They improve and become strong.'

Thus, the desire-energy was repeatedly channelized into the bodies, space, positions, objects and so forth which were more intensively tied to the market machine and reinvested in them. The desire to connect and place themselves closer to the market happened through creating a café space by working and improving their proximity to the market. Women embodied the logics of efficiency and maximization of profit by desiring to engage with the café as part of their own extended selves. The managers, through training and orientation, mentioned and constructed an enterprise which was to be embodied by the participating women foodworkers. This emerging enterprise was expected to create an affective ownership by the workers that further channelized the desire to engage with the café as "women who become stronger" through the café: "Devi had told me that this is a place to be built with our energy, seen to be our own, not somebody else's organization. We build it." "We built it together in the beginning, I spent all my energy. When I was asked to shift to the bakery, I wasn't even able to imagine that, I was being distanced from something that was part of me," said Vijaya after she left the café.

The women connected with the café space by re-territorializing themselves as part of the café. Their desire to earn and to develop themselves to be capable of earning was connected with many other desires and spaces, in multiple ways: first, with the desire of the managers who were closer to the market through their positions and qualifications and considered as hierarchically above the workers; second, the desire to connect with the market machine made them engage with these managers aspirationally. The desire to earn was thus reconstituted and reinvested as a desire to keep improving oneself, and to become a continuously improving and reskilling entrepreneur.

'Talking to people in a nice way, without being rude is helpful. See, Vimala cannot do that, she loses her temper and that often creates an obstacle for her to move on as a co-worker. As (better) workers, we have to be sweet and understanding too, only cooking is not enough here'

Thus, market-favoured behaviours were categorized as expected and accepted and market-unfriendly behaviours were categorized as unacceptable. Embodiment of such fine-grained codes and the attendant exclusions created conflicts and pressure for the participants. For example, Maya, the manager, talked about her not being able to ask the customers to leave the table when necessary, as such a talk is deemed to be unacceptable behaviour in the marketized café space: "When customers spend hours at our tables without ordering much, I cannot fight with them at the tables, in public. I've to be patient. No idea how to deal with such (stressful) situations."

Such stresses arose among workers as well, when they found themselves in conflict with each other's desires, which individualized them and connected their engagements competitively and more intensively with the market machine. They saw themselves as individuals, aggregated to increase their individual gains, improving themselves through creating more efficient and productive processes. This intensity led to blaming, experiences of fear and insecurities, "She is not competent enough for this role, makes too many careless mistakes. We should have someone competent to deal with the computers," says Rama about Prema. Prema, on another occasion, shares the discomfort of being judged negatively by Rama and the fear of losing her job:

'I know how to do my tasks. I do them very sincerely. I've often felt like resigning and finding a new job because of these kinds of experiences, but it is not easy to get a similar opportunity, moreover, Maya [Rama reports to Maya] treats me better, so I report directly to her'.

This bypassing of Rama, to whom Prema reported, and the sense of individualized competition recreated hierarchy. Here, certain engagements like those with the manager (Maya), who was placed closer to the market space through her role, skills and assigned tasks, and hence positioned above others in the emerging hierarchy, were considered more important. Conversely, the workers placed themselves below the manager's position, accepting the authoritative domination that adheres to the coding (that is, manager) and valuing the ability to control and coordinate the activities. Workers imagined the manager's positions to be closer to the market and they desired to connect with this position as such desirous engagements enabled them to further engage with the market. This placement was performed spatially when the manager occupied a position in the dining hall where the consumption of the foodwork happened, placing her closest to the market via direct customer interactions.

Yet there were sporadic moments in the café which escaped the frequent and intensive connecting of the desirous bodies with the market machine as Deleuze and Guattari argue. For example, workers connected and engaged with each other beyond the marketized desires, through other desiring machines, during their lunch time. They shared their food brought from home, explained how they make it while having lunch. Also, they shared their insecurities and vulnerabilities while eating lunch, as they were re-territorializing themselves into a friendly, collective space where they shared skills, interests, moments of happiness and celebrations and vulnerabilities.

'I was not able to cook today, but she will give me curry. I will eat chapatis with her curry. This happens sometimes, you know, we all have so much work to be managed at home as well, and we have to keep working all the time.'

'I wore a pair of jeans on last Sunday, see (showing the photo), for the first time. I was first very shy, but later I enjoyed.'

'I am not able to see my kids and family for so long now, they are far away from here, right, but we do not get jobs there. I tell myself, I am here for their good, earning for their education. So that is okay.'

'These sweets are made in a special style, you might not have eaten this, please try. This is not sold in any shop, this is made by our community, today is our festival you know'

Such instances of sharing and caring emerged when machines of desire engaged with non-marketized intentions. These instances of territorialization, rather than valuing proximity to the market, engaged with creating a collective space. In this way, affective solidarity (Vachhani and Pullen, 2019) was emerging and shaping the ways in which women interacted with each other. They cared, shared and supported each other, outside the affect of neoliberal organization of their bodies essentialized as 'women workers' in the larger process of production.

Conclusion

In this chapter, we have delineated the process of continuously territorializing and de-territorializing foodwork, workers and their gendered bodies into a café machine, where they come together not only to produce food but also to become part of the machine producing an affective hierarchy. The hierarchy emerges as desire flows through womanized working bodies, connecting them with other bodies that are considered better and higher in that they are closer to the market and embody market logics more intensively. Women workers embodied this positionality of being 'women' and 'worker' and desired to engage with others who embody similar positionalities, like those of women managers, thereby producing the café, a territorialized space. Through our empirical analysis, we have shown de/re-territorialization of women's bodies as 'bread earners' for the family and as entrepreneurs who continuously desire to engage with other production machines, creating affect that intensifies these desires to engage with the machines. Affect contributes to the production of human bodies as women, and thus to gendering the human bodies. These women improvise their actions in order to become better and/or improvise their positionalities to become empowered managers, responsible for the upliftment of their lower-rung co-workers. Womanhood, a collective affect that is experienced and essentialized by the parent organization, workers and their managers, emerges as the desire to engage with others who occupy similar positionalities. It shapes the desiring machines, re-channelizing the flow of desire in a specifically market-oriented direction, by territorializing the workers' bodies as women who need to mutually engage and entrepreneurially develop their positionalities to be closer to the market. Thus, neoliberal forces tend to become dominant in the café space by territorializing the worker bodies and creating an affective hierarchy, which would channelize the desires to engage with each other only in a marketized way, creating positionalities in a hierarchical order and denying the other possibilities of engagements. However, within the kitchen space, there are also sporadic but intense desires to connect with the bodies who are engaging in affective solidarity movements (Vachhani and Pullen, 2019), connections and embodiment through sharing joy, precarity and support, which militate against embodiment of market logics. The affective solidarity creates alternative ways of engaging, other than the affective hierarchical ways of engaging with the café space.

This chapter's contribution is twofold. First, we have shown the spatial emergence of a café as a production machine through connecting the desiring machines of women, an affective force that brings emergent engagements together. The café space under production is thus simultaneously a part of the production machine as well as an affective force that channelizes the desires and bodies via de/re-territorialization. In our narrative, the space itself becomes affective, as it shapes the engagements of different bodies when they create a production machine called a café. Further, affect becomes embodied through the positionalities that are created and performed by working bodies. Thus, we extend the discourse on production and experiences of affect through material embodiment (Beyes and De Cock, 2017) and describe its affective ability to control the flow of desire through space. Second, we have shown the de/re-territorializations that emerge within the affective space, where affective resistance (Baxter, 2021) against being limited and bound to one kind of territorialization – aligning to market interests – emerges. This affective resistance questions, reinterprets and reorganizes various engagements of desiring machines. Affective resistance is embodied in many forms: like in arguments between workers and managers and in precarities produced through affective hierarchy. Thus, we have shown that the territorialization process happens through these engagements and disengagements that occur in everyday life, in the embodiments and resistance of the affective hierarchization and in the secessions and disintegrations that occur within the café space. The processes that de-territorialize the female working bodies from the desire machine of hierarchized marketized production, as we showed, can pave the way to re-territorialize these machines into creating an affective solidarity. There the female bodies connect as transient and contingent machines that bring togetherness and allow the expressions of dissonance between the desiring machines of capitalistic production and the desiring machines of empowerment. Thus we have contributed to the debate on affective solidarity (Vachhani and Pullen, 2019) by showing the ways in which affective solidarity emerges through (de/re-)territorialization of working bodies while they experience and express affective dissonance and respond to it through creating local assemblages.

Note
[1] A plate (thali is a Hindi word for plate) of food with rotis or Indian bread, rice, a few
kinds of curries, sweet dishes and a few other dishes, depending upon which region of
the country it is related to. A North Indian thali, for example, will include rotis, and a
South Indian thali might include rotis and dal (a dish made of pulses and served with
roti) and have typical sweet dishes made in southern states of the country.

References

Acker, J. (1990) Hierarchies, Jobs, Bodies: A Theory of Gendered
Organizations. *Gender and Society*, 4(2): 139–158.

Aronson, J. and Neysmith, S. M. (1996) "You're not just in there to do
the work" Depersonalizing Policies and the Exploitation of Home Care
Workers. *Labor. Gender and Society*, 10(1): 59–77.

Baxter, L. F. (2021) The Importance of Vibrant Materialities in Transforming
Affective Dissonance into Affective Solidarity: How the Countess Ablaze
Organized the Tits Out Collective. *Gender, Work and Organization*,
28(3): 898–916.

Beyes, T. and De Cock, C. (2017) Adorno's Grey, Taussig's Blue: Colour,
Organization and Critical Affect. *Organization*, 24(1): 59–78.

Beyes, T. and Steyaert, C. (2021) Unsettling Bodies of Knowledge: Walking
as a Pedagogy of Affect. *Management Learning*, 52(2): 224–242.

Brenton, J. (2017) The Limits of Intensive Feeding: Maternal Foodwork at
the Intersections of Race, Class, and Gender. *Sociology of Health & Illness*,
39(6): 863–877.

Bhattacharya, T. (ed) (2017) *Social Reproduction Theory: Remapping Class,
Recentering Oppression*. London: Pluto Press.

Brenton, J. (2017) The Limits of Intensive Feeding: Maternal Foodwork at
the Intersections of Race, Class, and Gender. *Sociology of Health and Illness*,
39(6): 863–877.

Deleuze, G. and Guattari, F. (1988) *A Thousand Plateaus: Capitalism and
Schizophrenia*. London: Bloomsbury Publishing.

Deleuze, G. and Guattari, F. (2009) *Anti-Oedipus: Capitalism and Schizophrenia*.
London: Penguin.

Hemmings, C. (2012) Affective Solidarity: Feminist Reflexivity and Political
Transformation. *Feminist Theory*, 13(2): 147–161.

Hickel, J. (2014) The 'Girl Effect': Liberalism, Empowerment and the
Contradictions of Development. *Third World Quarterly*, 35(8): 1355–1373.

Lazzarato, M. (2014) *Signs and Machines: Capitalism and the Production of
Subjectivity*. Los Angeles: Semiotext(e).

Pedersen, M. and Kristensen, A. R. (2019) 'Blowing up the Pylon': the
Limitations to Lacanism in Organization Studies, Seen from the Perspective
of Deleuze and Guattari. *Culture and Organization*, 25(3): 189–201.

Ray, K. (2016) *The Ethnic Restaurateur*. London: Bloomsbury Publishing.

Ringrose, J. and Coleman, R. (2013) Looking and Desiring Machines: A Feminist Deleuzian Mapping of Bodies and Affects. In: R. Coleman and J. Ringrose (eds) *Deleuze and Research Methodologies*, pp 125–144.

Vachhani, S. J. (2015) Organizing Love – Thoughts on the Transformative and Activist Potential of Feminine Writing. *Gender, Work and Organization*, 22(2): 148–162.

Vachhani, S. J. and Pullen, A. (2019) Ethics, Politics and Feminist Organizing: Writing Feminist Infrapolitics and Affective Solidarity into Everyday Sexism. *Human Relations*, 72(1): 23–47.

PART II

Gender, Disruption and Unsettling Spaces and Places

7

Taking Place in-as Soho: Understanding the 'Here and There, Then and Now' of Gender and Affect Work

Melissa Tyler

Introduction

Affect theory provides a valuable lens through which to approach how 'placing' work involves grasping how the meanings and materialities that constitute particular spaces, places and settings are woven together into semblances of meaning and materiality (Simpson, 2021). Through an approach that draws on insights from affect theory and phenomenological geography, this chapter explores how working communities are spatially and temporally situated in ways that enable distinctive settings to 'take place' in two related ways. They do so first, by being affectively evoked as meaningful locations and second, by seizing particular settings through a distinctive, yet dynamic and evolving set of associations, imbuing them with a definitive set of expectations and capacities that shape how people perceive and relate to them, and to each other, in and through them. The chapter shows how, in the case of Soho, these are both hegemonically masculine and hyper-heteronormative, yet at the same time they are also critically queer and gender multiplicitous. The chapter highlights how Soho is a setting that brings together elements of its multiple pasts, presents and futures (its 'then and now'), and of reference points within and beyond its physical and perceptual boundaries (its 'here and there'), into a series of affective associations that, in combination, provide opportunities for gender to 'take place' in complex, contradictory and often also critically reflexive ways.

Structure-wise, it begins by considering working communities as affective places, drawing on insights and ideas from affective geography, phenomenology, and work and organization studies. It then goes on to examine Soho as a distinctive working community, considering its history and geography. It then considers Soho as a hyper-heteronormative, hegemonically masculine space, examining how this aspect of its affective atmosphere is encoded, embedded and enacted into its meanings and materialities, *at the same time and in the same space* as its being a place that opens up possibilities for queer politics and gender multiplicity to emerge. The chapter concludes by reflecting on how Soho retains elements of its 'edgy' history and distinct geography not in spite of but because of these tensions and juxtapositions that shape the ways in which Soho brings together place, gender and affect in enduring and distinctive ways.

Working communities as affective places

A growing interest in space, place and setting in work and organization studies has highlighted how the aesthetics and atmospheres of distinct places are important to understanding how working communities are constituted and experienced, both now and historically. As a theoretical 'way in' to studying working communities as 'meaningful locations' (Cresswell, 2004: 7), affect theory provides a valuable starting point through which to foreground the atmospheres that make particular places distinctive. It provides a lens through which to focus on how places come to exist as assemblages of affective associations and, further, how 'placing' work as both a social and analytical process involves grasping how the meanings and materialities that constitute particular places come to be woven together into affective semblances. Further, affect theory highlights how the latter give meaning to particular locales through the ways in which they 'take place' by simultaneously unfolding in distinctive settings and seizing those settings, such that a place and its working community may become indistinctive. The City, used to refer interchangeably to London's financial district and the UK's finance and banking sector, is an example (Nash, 2022, see also Simpson, 2021).

Edward Casey's (2001: 684) notion that there can be 'no place without self, and no self without place' sums up an affective understanding of place as the lived, felt and situated experience that constitutes distinctive, meaningful locations. It is a view that has gone on to inform research on place as 'an affective and relational compact' (Duff, 2010: 886), foregrounding affect as a radical openness that 'necessarily implicates bodies in ethical relations', and underpinning an approach to studying place that is, at least in part, concerned with a political-ethical critique of 'affective manipulation' (Pile, 2010: 8 and 17). This largely phenomenological way of thinking about place as an affective assemblage has evolved into a body of research emphasizing how 'humans layer their own understandings onto abstract space in order to create subjective places' (Jones and Evans, 2012: 2319) that are lived and experienced in situated, sensory ways that give meaning to particular locales through the affective associations invested in them. This growing body of research has highlighted the importance of 'geographies of affect' (Thien, 2005) in transforming spaces into places, with places being understood variously as 'pauses' in the rhythmic flow of space (Tuan, 1977), and as a constantly evolving set of social relations brought together at a particular location, with 'multiple identities that shift and overlap creating conflict and richness' (Massey, 1994: 153). And this body work, one that foregrounds affect, is an ideal starting point for studying communities like Soho, because within it place becomes both a geographic and temporal term, or, rather, a relation between past and future that is understood to form a place's distinctive 'social narrative' (O'Neill, 2007). From an affective perspective, place is, at once: 'The buildings, streets, monuments, and open spaces assembled at a certain geographic spot *and* actors' interpretations, representations, and identifications. Both domains (the material and the interpretative, the physical and the semiotic) work autonomously *and* in a mutually dependent way' (Gieryn, 2000: 467, original emphasis).

This means that a sense of place is not simply the ability to locate things cognitively, on a map for instance. It is also the attribution of meaning and sensual attachment to a particular location – spaces become places when they are physically, materially invested with affective, sensory associations. Places are imbued with associations that are not simply connected to the people or organizations that occupy them; the controlling, compelling or constraining effects of particular places emerge from the location, built environment and symbolic sense we get of specific settings, so that the material and the social are mutually influential (Massey, 1997). Places are spaces *that mean*, and at the same time, they are meanings that are *materialized* in discernible locales; they are the settings that make us feel something in or about them.

To understand this, Casey (2001: 684) distinguishes between what he calls thick and thin places in order to highlight the role that affect plays in the production of place. For him, thin places are those that increasingly characterize over-developed settings in contemporary, neoliberal economies. They are places that have been erased of any local specificity, and which become devoid of any unique quality or feature. Orientated primarily towards commercial accumulation, they are what he calls 'levelled down' settings whose resemblance to a myriad of others like it makes them instantly familiar and easily navigable, especially for consumers. In contrast, 'thick' places are those with and through which we develop a close association, and in which, for Casey, we can flourish. Duff (2010: 886) notes how any city is made up of thick and thin places, equating these (respectively) to 'places which support diverse enriching experiences, and places which leave no trace at all'. But cities and other meaningful locations are also made up of abject spaces – places which are 'thick' in their affective atmospheres and associations, but which do not necessarily support enriching experiences or nurture close affinities. Rather, abject spaces might be thick in the ways in which they are replete with meaning, but which are perceived and experienced as simultaneously seductive and repulsive in complex and often apparently contradictory ways, as illustrated by Nash's (2022) study of women working in the City of London, noted earlier. And these are quite distinct from the ways in which Duff (2010: 893), drawing on Casey (2001), understands thick places when he says, 'to feel connected to place is to experience a sense of belonging in place that itself generates resources of immense value in the promotion of health and well-being'. Places which are 'thick' in Casey's terms are dense and heavy with meaning, but they can equally be those which generate a sense of abjection rather than affiliation, and London's Soho is arguably just such a place.

Duff (2010) describes how the affective production of thick places can take the form of a series of 'relays', showing how the atmospheres created by these relays help to transform what Casey describes as 'thin', meaningless spaces into 'thick places', replete with significance. Duff notes how to experience place is to be affected by it, a view that echoes Anderson's (2009: 78) notion of places as settings that are generative of 'affective atmospheres'. For Duff (2010: 881–2, emphasis added), affect refers to 'the lived sensation, the feel, and emotional resonance of place', so that affective atmospheres capture not just 'the emotional feel of a place' but also a sense of what is 'potentially *enactable* in that place'.

Again, this is a pertinent point when applied to Soho, which has a reputation as being an 'anything goes' kind of place (Tyler, 2020). And it is important to keep in mind, in this sense, that affect theory and efforts to incorporate it into phenomenological geography rest not simply on an appreciation of the role of emotion in shaping human experience, but also on a grasp of the ways in which human interaction, capacity and 'potential' is transformed by that experience. As Duff (2010) notes, at least since the time of Spinoza, affect theorists have argued that affect conveys much more than a ceaseless transition from one feeling state to another. Spinoza (1989 [1677]: 129–130) contended that affects constitute the body's 'power of action', that is, its unique capacity to affect (and be affected by) the other bodies and things that it encounters. Contemporary interest in affect picks up on this, highlighting how affect constitutes an embodied experience 'that has the capacity to transform as well as exceed social subjection' (Hemmings, 2005: 549). Queer theorists have developed this view, showing how shame can induce empathetic resonances beyond its homophobic intent, creating opportunities for affect-bound communities and solidarity to emerge

(Burchiellaro, 2021a, 2021b); again, a point that resonates with a place like Soho, as we discuss in more detail later.

For now, it is important to note that this way of thinking about affect helps us to understand how places are made, enabling us to grasp not only how they are constituted in ways that carry distinctive emotional associations but which, as they are worked into thick, meaningful locations, open up distinctive 'capacities'. In other words, an affective approach brings to the fore how place making is not simply a one-way process through which a location is transformed into something or somewhere distinctive, it highlights how places also 'make' what goes on in and around them. Understood in this way, places become spaces of expectation, settings that have particular affective associations, through which those who experience them are compelled or constrained, or imbued with the capacity to act in certain ways. It follows, then, that affective encounters involve embodied experiences that extend well beyond the individual and take place within and through social relations, and that these two aspects of affect, the feeling states generated in, by and *as* places, as well as the capacities, practices and dispositions that distinctive places make possible, give form to them as affective atmospheres. Once these become 'thickly' (Casey, 2001) encoded, enacted and embedded into the complex and dynamic constellation of meanings and materialities that make places, and that places make, spaces are transformed into meaningful locations pregnant with possibilities as affective communities of recognition and solidarity, including, for example, as settings for queer solidarity and gender multiplicity.

Soho as a meaningful location: its 'then, now, here and there'

Tuan's (1977) notion of place as a pause in the flow of space is particularly apt for Soho. With its squares, benches and cafes and long history of migration, Soho has throughout its history been thought of as a place where the wider flows of time, capital and people 'pause', whether fleeing from religious or political persecution en masse, or simply taking some time out from travelling in a gap year. One of the risks associated with thinking about place in this rather romantic way, however, is that it seems relatively static and stable, in contrast to the dynamic and constant 'flow' of space. However, as somewhere like Soho, and London more generally, aptly illustrates, the materialities and meanings, assemblages and associations that constitute particular places are constantly changing. Place is more of a process than a fixed point, a change in the pace and tempo of space rather than a distinct pause, or a sensory multiplicity experienced rhythmically, as Simmel (1964: 19) describes it. Alec Waugh (1920: 136) hints at this in his account of Soho, written in the 1920s, when he notes the place's distinctive transience, 'where life is fleeting and uncertain' and new atmospheres are quickly created and recreated as those who make up Soho's working community are in a 'constant state of becoming'. Waugh is referring to the constant flow of restaurant cooks, bar and waiting staff who work there, but he could just as easily have been making a point about the ontology of the place itself, constituted as it is through a complex process or assemblage of affective entities and associations.

Described by some as 'the most tempting and exciting part of London' (Hutton, 2012: 7), Soho is a relatively compact geographical space, with a distinct local geography and character; yet in terms of its wider cultural associations and economic significance it is also a highly global setting. Despite recent concerns about corporate over-development and sanitization (Sanders-McDonagh et al, 2016), Soho retains an 'edge' to it that reflects its history, culture, politics and location, and perhaps most notably, its continuing association with the commercial sex industry (Speiser, 2017; Tyler, 2020). A comparatively transient place embedded within a history of migration and cultural diversity, Soho's economy is characterized by sectors of work that are relatively unstable and precarious. At the same time, it is within close proximity to London's main shopping thoroughfares, Oxford Street and Regent Street, and to the more upmarket area, Mayfair. The latter is a key location for private equity and investment banking firms, both of which feed off Soho's long-standing association with enjoyment and entertainment, exploitation and excess (Glinert, 2007; Riach and Tyler, 2022).

As a place that 'plays to all the senses' (Massey, 2002: 458), Soho has, throughout its history, been something of an abject space, maintaining a long-standing appeal as simultaneously alluring and threatening, exploiting many of those who work and consume there, at the same time as carefully nurturing its reputation as a place of bohemian indulgence, offering a warm embrace and a sense of belonging in the heart of an otherwise relatively anonymous urban environment (Tyler, 2020). The area's renowned pleasure economies in food, theatre, fashion, music and commercial sex intersect with its built environment, material culture and distinctive geography. Added to this, Soho's history of migration, its propensity for reinvention, eclecticism, political and religious diversity, and the setting's distinct location and urban geography, combine to enable Soho to 'take place' in a very particular way. This 'taking place' both shapes the work experiences and identities of those who work there, and 'work' Soho into being the place that it is, bringing it into being through the associations and expectations attached to it to situate or 'place' lived experiences, identities and perceptions. A place of complexity and contrast, Soho's most recent renaissance means that it is now associated as much with high-specification information and communications technology and post-production media, and a vibrant restaurant and bar industry, as it is with commercial sex. Urban branding, local community initiatives and the introduction and enforcement of licensing regulations have combined to 'clean up' Soho, as the twin processes of gentrification and corporatization have, some would argue, sanitized the area beyond recognition. Yet, as a place to live, work and consume, Soho retains an 'edge' to it that can still be discerned only just beneath its increasingly corporate surface, or 'thin' veneer, to borrow from Casey (2001).

Soho's provenance as the centre of London's cosmopolitan pleasure zone dates back at least to the early 18th century. The area's proximity to political and legal power bases, and to the retail and entertainment heart of the capital, created a seductive atmosphere that brought migrants, tourists and visitors in from across the world. This has defined Soho throughout its history and continues to do so today as its history intersects not only with the area's location but also with Soho's distinctive geography and the materiality of its built environment, and these patterns of social and physical space shape Soho as a distinctive workplace. Soho's symbolic and social geography, its location and accumulated history, have all played a significant role in imbuing it with a distinctive atmosphere constituted through the area's social materiality.

Soho's history and economy, its geography, culture and industry are much more fluid and multi-dimensional than surrounding areas of central London that are dominated by largely corporate commercial, political and financial organizations. There is a well-established association with the creative industries there, notably post-production film and music, and the expansion of businesses specializing in digital media, communication technologies and financial services has added to these aspects of the area as a working community since the 1990s. However, the area and its economy continue to be shaped largely by small-scale, local businesses, and by clustered markets, most notably in hospitality and commercial sex. With regard to the latter, while urban branding, local community initiatives and the introduction and enforcement of licensing regulations since the 1980s have combined to 'clean up' Soho, the continuing presence of commercial sex and an enduring association with sleaze makes it a distinctive place to live, work and consume. And although Soho is a continually evolving locale, the area's historically established associations and enduring uniqueness continue to shape perceptions of what it means, and of how it is experienced, as a working community today.

Understanding this helps us to think about how particular kinds of work or industry 'take place' in Soho, if taking place refers both to assuming a place (taking 'possession' of it) and in some sense to happening and occurring, to coming to exist in a recognizable form within the context of that place. Placing, in this sense, is a situated process of social and material enactment that evokes meaningful locations through the attribution of particular affective associations; for example, the commercial sex industry 'takes place' in and through Soho insofar as it simultaneously happens in, and dominates, the setting. The materialities that bring this into being include its streets, courts and alleyways; its signage and other semiotic artefacts; the fronts and entrances of its buildings, including shops, clubs and other venues; the interiors of these settings, including their layout, lighting, 'front' and 'back stage' areas, their props and product ranges, as well as the physical presence of those who live, work, consume, pass through or otherwise 'hang out' in the area.

Together, although not necessarily in any harmonious way, these materialities mean that Soho 'takes place' through its physical and embodied substance, and the meanings attributed to these – the social, cultural, political and affective associations with the aforementioned substance that brings Soho into being in a very *particular* way. The latter includes the area's accumulated historical and contemporary associations, or what Simmel (1964: 11) describes as 'the weight of historical heritage', with bohemianism, artistry and community, and with sleaze, exploitation and over-development, as well as sanitization and commercialization. Taken together, these materialities and meanings constitute Soho as a distinctive working community that 'takes place' and in which, as a significant part of what Soho is as an affective assemblage, gender dynamics are played out.

Soho as a 'thickly' gendered location

Relating gender to place and setting is by no means new, since this association follows from an ontology of gender as a largely situated social process rather than a 'property' of the self. As Casey (1993: xiii) has put it, the places we occupy have 'everything to do with what and who we are', and Soho is a setting in which gender relations and identities are played out in ways that are significant to what the place *is* as a nexus of meanings and materialities. Gender is 'thickly', to borrow from Casey (2001), encoded, enacted and embedded in complex and evolving ways into what Soho means, as an assemblage of affective associations. At one level, these 'serve both capital and the phallus', as Hubbard (2004: 670) puts it with reference to hyper-masculine city-centre playscapes. Yet, Soho is also a distinctively queer setting, as is discernible in the sensations and associations conjured up by its aesthetic landscape, and by the affective capacity the area evokes for ways of being and interacting that actively challenge and resists heteronormativity and hyper-masculinity.

Soho as a place of hegemonic masculinity and hyper-heteronormativity

As a place in which commercial sex is concentrated in highly reified forms, for example, in its clubs, cinemas, shops, bars and massage parlours, it is not surprising that Soho has a long and enduring association with a hyper-masculine consumption of women's and men's bodies, and with gender appropriation. Its contemporary semiotic landscape is replete with hyper-heteronormative depictions of sexual objectification and its signage attests to continued commercial exploitation. And it is no coincidence that the men being trained as pick-up artists practise the predatory techniques they have learnt in the classroom in field-based 'gaming' exercises in and around Soho's streets and bars (O'Neill, 2018). Soho began as a hunting ground during the reign of Henry VIII and remains so today.

Historically sedimented perceptions of Soho as a male-only space, or as 'rightfully' a male-dominated place, have a long history. Indeed, books depicting life in 19th- and early 20th-century Soho such as Stevenson's *Jekyll and Hyde*, Ransome's *Bohemia in London* and Conrad's *The Station Agent* contributed to an enduring literary myth that framed Soho as a hegemonically masculine setting. Other commentators such as Daniel Farson in the 1950s, and biographers of Soho sex-industry entrepreneurs such as Paul Raymond and Murray Goldstein, as well as accounts of the vice squad and criminal gangs throughout the second half of the 20th century (Tyler, 2020), all contributed to sustaining this affective association, one that is materialized in contemporary Soho's semiotic landscape.

Women have played a vital part in Soho's cultural and commercial history, however, but largely as those who are consumed rather than as active consumers; women are the objects rather than the subjects of its narrative. Speiser (2017: 18) describes how Edwardian Soho was 'the world's largest flesh market, its streets after dark almost entirely given over to sexual commerce'. Despite a contraction of the sex industry in the hundred years or so since then, and the movement of sex workers off the streets and into flats and massage parlours following the introduction of the 1958 Street Offences Act, Soho today is not so different from this. The transactional nature of women's presence there is widely signified in the area's material and cultural landscape, replete as it is with sexualized displays of women's bodies as objects to be consumed.

Marketing and merchandizing efforts to attract more female consumers to Soho have arguably been used to regulate its sex industry and the area as a whole by reframing its hyper-masculine associations with the place's 'seedy' past. It is unlikely that such tactics will proffer anything akin to genuine challenges to the heteronormative hegemony that characterizes Soho's affective atmosphere, however, not least because they are articulated largely through a consumer-orientated discourse, framed in terms of rhetorical references to empowerment, liberation and autonomy. And at the same time, as noted earlier, the area's dominant semiotic landscape continues to reproduce objectifying imagery of women's bodies. Crew and Martin (2017: 593–4) emphasize how the commercialization and commodification of sexuality in places like Soho is 'both deeply classed and profoundly gendered', arguing that their commercial environments connect to discourses of postfeminism insofar as 'via pink stores and products … they make claims of sexual empowerment and yet continue to emphasize the female body as sexual object'. Conforming to the gender-normative terms of the heterosexual matrix (Butler, 2000), Soho materializes a heteronormative, hegemonically hyper-masculine version of what it means to be a woman that is signified, in its commercial sex industry, almost ubiquitously through the use of the colour pink.

At one level, this means that it would be very easy to present Soho as an anachronistic, phallocentric place. Yet, that would be an over-simplification that would fail to do justice to the complexity and multiplicity that characterizes the area as an assemblage of contrasts and contradictions that, now and historically, elude simple categorization. Soho's historical associations, geographical positioning as an urban village and its distinctive materiality, not to mention the area's long-established reputation as a community of outsiders, all mean that, while the commercial sex industry there continues to 'undo' those men and women who are caught up in it, it also constitutes a space in which sexual objectification, exploitation and constraint can potentially be 'undone' in ways that reflect its history as an eclectic community of outsiders, offering those who don't feel as if they belong elsewhere a chance to be 'at home'. Now and historically, as well as being a site of sexual exploitation *in extremis*, Soho has also been a place in which queer solidarity and gendered multiplicity can flourish, and a place where the gender conventions might be unravelled in order to be done differently.

Soho as a critically queer and multiplicitous place

Alongside the characteristics of Soho as an affective assemblage that 'takes place', discussed earlier, therefore, Soho is also somewhere that attracts those whose sexualities, points of gender identification and lifestyles position them as somehow 'outside' of the mainstream, with many of those who live, work and consume in Soho gravitating there as a queer space of belonging and community. In this sense, it constitutes a much 'thicker' place, in Casey's (2001) terms, than the description provided earlier might suggest. The tightly packed crowds taking part in a candlelight vigil on Old Compton Street the day after the mass shooting of LGBTQ people at the Pulse nightclub in Orlando, Florida in June 2016, and the extent to which Soho remains a focal point for London's annual Pride festivities, are just two illustrations of the sense of connection that the place provides at a collective level, just as an affective sense of gravitating towards and 'belonging' to Soho might do at a more individual level.

The area's well-established reputation as an oasis of Otherness means that Soho has, throughout its history, been a socially, culturally and politically important place in which alternative ways of doing gender and sexuality might be possible, and those who live outside of normative conventions might find recognition. In this sense, it is also a setting that provides some scope for challenging the normative expectations and exploitative associations already considered, and in which other ways of being might flourish. One of its more famous 18th-century residents was the Chevalier d'Eon, who moved relatively seamlessly between living as both a man and a woman and in doing so set the foundations for Soho's subsequent association with sex and gender as multiplicitous possibilities rather than fixed categories to flourish (Tyler, 2020).

Over recent decades, developments in ways of thinking about, understanding, representing and experiencing gender mean that a binary ontology of men and women as belonging to fixed, pre-social gender categories has been superseded, or at least supplemented by a more performative way of conceiving of gender as a process that is enacted within everyday social interactions, and which is continually renegotiated. Such an approach highlights that gender is less 'a property' possessed by individuals, or a category according to which men and women can be classified in binary, hierarchical terms. Instead, this performative approach emphasizes that gender is a verb – an ongoing process, often one that involves a struggle for recognition and to be treated with dignity and respect (Faye, 2022), enacted within social relations. Further, this approach emphasizes that this 'doing' of gender does not happen in a social vacuum; it is situated within the materiality, the combined history and geography, of particular social contexts, locations and settings. It is on the recitation of gender norms over time that Butler (2004) focuses much of her attention in her performative analysis of gender as a perpetual process of un/doing. But, as a socially and culturally situated process, gender is also performatively enacted and 'struggled' over within the context of specific settings and places.

As discussed earlier, the social materiality of Soho is heavily encoded with heteronormative, gendered imagery, *at the same time* as it opens up the possibility for this to be challenged. The area's burgeoning shops, bars, restaurants; its history and geography, and the 'anything goes' ethos that shapes Soho as a working community, all hint at this. If, as work settings, places 'do gender' rather than simply have it done to them, Soho is clearly a heavily masculine space, one that materializes the heterosexual matrix as Butler (2000) outlines it. Yet, at the same time it suggests both the consequences of this for 'unravelling' those subject to its normative ethos, *as well as* providing the material setting, an affective beckoning so to speak, for this normativity to be challenged, opening up the possibility of being and doing things differently. As already noted, Soho has a long history of making 'trouble' with fixed identity categories, including those shaping gender and sexuality. From the Chevalier d'Eon to Madame Jo Jo's renowned performance queens (as a tribute to whom the artist Christine and the Queens chose their name), queer and multiplicitous ways of being have always 'taken place' there. In contemporary Soho this is perhaps most obviously apparent in and around its central thoroughfare, Old Compton Street, where the presence of lifestyle stores and a proliferation of bars, restaurants and clubs put out the LGBTQ 'welcome' mat, and rainbow flags fly all year round, rather than being largely reserved for Pride, as they tend to be in other places.

In sum, Soho is a setting that has historically aligned itself with queerness, as a working community that problematizes categorization, and, in doing so, ostensibly disrupting hegemonic gender regimes in a place that constitutes Soho as an affectively queer constellation. Ostensibly signalling, or at least hinting at, an undoing of gender in Butler's (2004) terms, this aspect of Soho as a working community is one that contrasts with a view of the place as overwhelmingly, heternormatively hyper-masculine; produced for men, staffed predominantly by men and frequented largely by men (Tyler, 2012). It resonates instead with a sense that Soho also takes place as an affective atmosphere that is more fluid, and which potentially holds within its space the capacity for queer solidarity and gender multiplicity.

To link this to theoretical perspectives on gender a little more, emphasizing the importance of what she calls 'the scenography of production' in understanding the relationship between gender performativity, materiality and signification, Butler's (2004: 1) analysis of gender as a perpetual process of 'un/doing' highlights both the agentive capacity of the subject and the constraints that compel and/or restrict the affective possibilities attached to gender. In her discussion of the latter, she focuses specifically on the dialectical interplay between 'what it might mean to undo restrictively normative conceptions of sexual and gendered life' (Butler, 2004: 1) and the matrices of cultural intelligibility that shape the hegemonic performance of gender in ways that will be accorded recognition according to the terms of the heterosexual matrix. The latter involves a process that Butler describes as 'becoming undone', denoting the ways in which the complexity of lived experience becomes conflated, unravelled so to speak, in the construction of a self that is compelled or constrained in particular ways in order to become socially acceptable. The basis of her critique is that this perpetual process 'imposes a model of coherent gendered life that demeans the complex ways in which gendered lives are crafted and lived' (Butler, 2004: 5). The area's largely commercial orientation notwithstanding, what places like Soho potentially open up is the possibility of reinstating and resignifying at least some of this complexity, as a place in which to challenge even in a relatively minor way the terms of the heterosexual matrix, or 'presumptive heterosexuality' as Butler (2004: 186) put is. And it is the possibilities associated with being able to 'make trouble' with gender, to borrow further from Butler (2000), that, at least in part, explain the ongoing attraction of Soho as a working community.

Conclusion

While to say that Soho constitutes 'the epitome of hard-core hedonism' (Glinert, 2007: ix) might be overstating the case, Soho remains an important place for mobilizing queer solidarity and gender multiplicity; it retains an 'affective pull' towards which those in search of recognition and respite can gravitate. But in combination with its location, this sense of community and connection is also, at least in part, what makes Soho perpetually vulnerable to corporate over-development, to a 'thinning out' in Casey's (2001) terms. Yet, despite ongoing development plans driven by a mimetic desire to capitalize on Soho's edgy past that carry with them the risk of crushing its character in the process, creating the whole area as a sanitized simulacra much like other themed neighbourhoods or living museums in urban centres, Soho continues to have a global reputation for its 'edge'. In the heart of a global city, it continues to thrive on its atmosphere as an urban village and as a working community in which gender constitutes a range of possibilities. As a thriving, working community of outsiders, Soho is a space/site on which struggles for gender recognition are played out, and in which hegemonic hyper-masculinity and heteronormativity is built into its very fabric, but in which contestation and ongoing struggle with its pasts/presents, its 'heres' and 'theres' opens up scope for alternative ways of being and 'taking' place to emerge. As an affective assemblage of meanings and materialities, Soho is a setting in and through which the 'work' or capacity needed to un/do gender as an affective endeavour is played out in what seem to be continually evolving ways. Its 'feel' for gender hegemony and multiplicity sit side by side, not always in harmony, but placed in the same compact space and compressed into shared references to its pasts, presents and (imagined) futures.

References

Anderson, B. (2009) Affective Atmospheres. *Emotion, Space and Society*, 2: 77–81.

Burchiellaro, O. (2021a) 'There's Nowhere Wonky Left to Go': Gentrification, Queerness and Class Politics of Inclusion in (East) London. *Gender, Work and Organization*, 28(1): 24–38.

Burchiellaro, O. (2021b) Queering Control and Inclusion in the Contemporary LGBT-friendly Organization: On LGBT-friendly Control and the (Failed) Reproduction of (Queer) Value. *Organization Studies*, 42(5): 761–758.

Butler, J. (2000) *Gender Trouble*. London: Routledge.

Butler, J. (2004) *Undoing Gender*. London: Routledge.

Casey, E. (1993) *Getting Back into Place: Toward a Renewed Understanding of the Place-World*. Bloomington, IN: Indiana University Press.

Casey, E. (2001) Between Geography and Philosophy: What Does It Mean to Be in the Place-World? *Annals of the Association of American Geographers*, 91: 683–693.

Cresswell, T. (2004) *Place: A Short Introduction*. Oxford: Blackwell.

Crewe, L. and Martin, A. (2017) Sex and the City: Branding, Gender and the Commodification of Sex Consumption in Contemporary Retailing. *Urban Studies*, 54(3): 582–599.

Duff, C. (2010) On the Role of Affect and Practice in the Production of Place. *Environment and Planning D: Society and Space*, 28: 881–895.

Faye, S. (2022) *The Trans-gender Issue: An Argument for Justice*. London: Penguin.

Gieryn, T. F. (2000) A Space for Place in Sociology. *Annual Review of Sociology*, 26: 463–496.

Glinert, E. (2007) *West End Chronicles*. London: Allen Lane.

Hemmings, C. (2005) Invoking Affect. *Cultural Studies*, 19(5): 548–567.

Hubbard, P. (2004) Revenge and Injustice in the Neoliberal City: Uncovering Masculinist Agendas. *Antipode*, 36(4): 665–686.

Hutton, M. (2012) *The Story of Soho: The Windmill Years 1932–1964*. Stroud: Amberly.

Jones, P. and Evans, J. (2012) Rescue Geography: Place Making, Affect and Regeneration. *Urban Studies*, 49(11): 2315–2330.

Massey, D. (1994) *Space, Place and Gender*. Cambridge: Polity Press.

Massey, D. (1997) A Global Sense of Place. In: T. Barnes and D. Gregory (eds) *Reading Human Geography*. London: Arnold, pp 315–353.

Massey, D. (2002) Living in Wythenshawe. In: I. Borden (ed) *The Unknown City: Contesting Architecture and Social Space*. Cambridge, MA: MIT Press, pp 458–475.

Nash, L. (2022) *The Lived Experience of Work and City Rhythms*. Bingley: Emerald.

O'Neill, J. (2007) *Markets, Deliberation and Environment*. London: Routledge.

O'Neill, R. (2018) *Seduction: Men, Masculinity and Mediated Intimacy*. Cambridge: Polity.

Pile, S. (2010) Emotions and Affect in Recent Human Geography. *Transactions of the Institute of British Geographers*, 35: 5–20.

Riach, K. and Tyler, M. (2022) 'Getting a Grip'? Phenomenological Insights into Handling Work Place in London's Soho. *Human Relations*. https://journals.sagepub.com/doi/pdf/10.1177/00187267221135016

Sanders-McDonagh, E., Peyrefitte, M. and Ryalls, M. (2016) Sanitising the City: Exploring Hegemonic Gentrification in London's Soho. *Sociological Research Online*, 21(3).

Simmel, G. (1964) The Metropolis and Mental Life. In: K. Wolff (ed) *The Sociology of Georg Simmel*. New York: The Free Press of Glencoe.

Simpson, A. (2021) *Harm Production and the Moral Dislocation of Finance in the City of London: An Ethnography*. Bingley: Emerald.

Speiser, P. (2017) *Soho: The Heart of Bohemian London*. London: The British Library.

Spinoza, B. (1989) *The Ethics*. New York: Prometheus Books.

Thien, D. (2005) After or Beyond Feeling? A Consideration of Affect and Emotion in Geography. *Area*, 37(4): 450–454.

Tuan, Y.-F. (1977) *Space and Place: The Perspective of Experience*. London: Edward Arnold.

Tyler, M. (2012) 'Glamour Girls, Macho Men and Everything in between': Un/doing Gender and Dirty in Soho's Sex Shops. In: R. Simpson, N. Slutskaya, P. Lewis and H. Höpfl (eds) *Dirty Work*. London: Palgrave Macmillan, pp 65–90.

Tyler, M. (2020) *Soho at Work: Place and Pleasure in Contemporary London*. Cambridge: Cambridge University Press.

Waugh, A. (1920) Round about Soho. In: S. J. Adcock (ed) *Wonderful London: Volume One*. London: The Fleetaway House, pp 129–136.

8

Affective Practices and Liminal Space-making in Palestinian Refugee Camps

Alison Hirst and Christina Schwabenland

Introduction

What is the role of affect in making places? What is its role in making or changing borders and the places they separate and connect? We consider these questions in the context of two Palestinian refugee camps: Bourj al Barajneh in Beirut, Lebanon, and Aida Camp in the Occupied Territories, just outside Bethlehem. These two camps number among many that were established in 1948–49 in response to the displacement of approximately 700,000 Palestinians who fled their homes and villages during the 'Naqba', or 'catastrophe', and who have been given, to date, no right of return. Although the word 'camp' suggests impermanence, the original tents have been replaced with storeys-high breezeblock constructions. Hanafi et al (2012: 39) term the camps 'enclaves of exception', reflecting their indeterminate status: camp residents do not have the rights of full citizens, but 'refugee' seems an inappropriate term for people whose families have been living in the camps for such a long time.

The absence of a political settlement has left the original refugees (those who are still alive) and their descendants living in very poor conditions, with poverty and unemployment endemic. Both of the camps are easily distinguishable from the neighbouring areas due to the poor quality of the buildings and roads, and their high-density construction, with buildings covering all but very narrow roads and walkways. Bourj al Barajneh is more densely settled and the Lebanese government provides no municipal services; water and electricity are 'illegally' brought into the camp, the latter circulated via festoons of overhead cabling. There are just two entrances/exits, and those entering and leaving are observed via local militia stationed at the entrance. Aida Camp, by contrast, is more spacious and a little better resourced, with electricity and water provided by the Palestinian Authority. However, Aida Camp is located adjacent to the Israeli Barrier Wall, a menacing construction made from huge concrete slabs interspersed with watchtowers that enable constant surveillance. For women, the ongoing effects are that their daily lives are narrowly lived, confined both within the camp boundary and also within a second, less visible boundary between their homes and the wider camp. Peteet (2017) states that enclosure can lead to reinforcement of traditional patriarchal gender roles, but it can also provide opportunities for women's empowerment.

Berlant's (2010) complex and bittersweet epithet 'cruel optimism' refers to an individual's or group's enduring attachment to some object of desire that is unlikely ever to materialize. Such optimism can endure, says Berlant (2010), because the subject's sense of self and hope for the future are lodged within it, and conversely, its loss 'will defeat the capacity to have any hope about any thing' (Berlant, 2010: 94). The hope of return is symbolized for many Palestinian refugees by the metaphor of the key, representing the doors to the homes they had to leave and to which they desire to return. A gigantic model of the key overhangs the entrance to Aida camp (Figure 8.1). Many Palestinians also have more modest hopes, for 'normal lives, the absence of disorder, some measure of predictability, and the sense of dignity that comes with recognition of one's humanity' (Peteet, 2017: 202).

Within this context of indefinite refugee status, confinement and exclusion, and unreasonable hope, we consider a kitchen (established in 2014) and two rooftop gardens (each established in 2016). Kitchens and gardens are generally domestic, humble settings, not usually considered as spaces that can complicate gendered identities or enable more choice, mobility and freedom for women to decide who they are and would like to become. In these cases, however, the kitchen and the gardens have created openings through the home/camp and camp/outside borders, enabling the women to leave their homes, to work, as workers to leave the camp to sell food, and also as workers to become hosts who welcome guests and offer hospitality to others. In the case of one of the rooftop gardens, the opening of the border also makes the garden vulnerable to repeated military violence, and it has been be remade and replanted again and again. In both cases, what were 'strong' socio-spatial borders have been disrupted, with openings created across them. They have acquired properties of liminal spaces, where the usual everyday social norms are lifted or complicated, and through which occupants can take up different roles and identities.

According to Massey (2005), any space (such as a boundary, wall or kitchen) 'is always in the process of being made. It is never finished, never closed' (Massey, 2005: 9). We use Wetherell's (2012) concept of *affective*

Figure 8.1: The 'Key of Return' at the entrance to Aida Refugee Camp

Source: Christina Schwabenland

practice to analyse how the borders of the camps have been redrawn, and the consequences for the women making and remaking these spaces. Affective practice highlights the doing of affects and their emergent consequences – doing by and through the active participation of assemblages of humans and non-humans with diverse ontologies (kitchen equipment, husbands, cookery 'know how', hydroponic techniques and seedlings, for instance). Wetherell (2012) also considers critically whether and how we attune to the affective practices we encounter, and whether or not we too choose to become participants in collective 'atmospheres'. This is important because the spatial boundaries of the camps we consider mark extreme social divisions.

As well as seeking deeper insights into the situation of Palestinian women refugees and support for their cause, we also aim for better understanding of how affective practices contribute to the construction of liminal spaces in which identities can be changed or complicated. A liminal space implies a threshold that can be crossed: it creates new openings. In this case, we find that the domestic spaces do not simply replace one set of affective practices with others. Rather, they combine public with private affective practices so that there is both continuity and improvisation at the same time, in ways that offer reassurance to insiders and outsiders.

Borders and liminal spaces

Borders

For Massey (2005: 68), space is an 'emergent product of relations, including those relations which establish boundaries'; boundaries segregate, and they also connect. Material boundaries solidify and make visible their configurations of separation and encounter, but, as Massey (2005: 93) writes, there are 'also "others" within: not least, though also not only, "women" and "nature"'. Taking up Massey's relational understanding of space, Sohn (2016: 184) argues that we should conceptualize bordering practices as assemblages composed from 'actors, objects, practices and representations'. This enables us to understand their multiplicity, 'the diversity of practices attached to [them] and that give strength to a multitude of specific versions' (Sohn, 2016: 184). What is significant, suggests Sohn (2016), is not only the diversity of versions but how they interact with one another, and how some responses are regarded as legitimate and others not.

Borders mark a 'limit between two territorial and social entities' (Sohn 2016: 184), and, in each of the camps we consider, the camp boundaries mark the outlines of what Peteet (2017: 61) terms 'warehouses for the excluded'. In Beirut, venturing outside the camp means encountering suspicion and prejudice from the Lebanese (Abu Mughli, 2020). In Bethlehem, although, as Massey (2005: 9) argues, space is 'never closed', the Separation Barrier is a forceful enclosure. Boyce et al (2015) comment on the different affective resonances that the wall has for the people it separates and connects. For Israeli citizens, they suggest, the wall '[assuages] the sense of insecurity caused by ill-defined boundaries. This allows for a collective psychological, if not physical, disengagement from the Occupied Territories' (Boyce et al, 2015: 289). For Palestinians, the wall 'deeply [embeds] Israeli military power within the terrain of Palestinian everyday life' (Boyce et al, 2015: 289). The wall severely limits all Palestinians' mobility, with the result that 'Palestinians are increasingly constrained in their place-making capacity. They can only craft and give meaning to place in very delimited areas' (Peteet, 2017: 137). Within both of the camps, women tend to confine themselves to their homes, primarily engaging in domestic activities; indeed, for camp women, domestic duties are 'the arenas where femininity was enacted and reaffirmed' (Peteet, 2005: 186). Thus, for women, the camps function simultaneously as a place of safety and a prison (Schwabenland and Hirst, 2022).

Liminal spaces

The creation of the kitchen and gardens we consider has brought about physical modifications to the camps' boundaries, and these modifications have enabled women to take up new roles and identities. To help understand how they bring about changes, not only to spatial borders but also to the gendered identities of the women who make and use them, we have conceptualized them as liminal spaces. To go back one step, liminal experiences take their name from *limen*, the Latin word for threshold, and are the critical stage in 'rites of passage' that bring about the transformation of identity of the liminal sunject (van Gennep, 1960; Turner, 1974). Such experiences can be transformative, because in liminality people are released from everyday norms. Turner (1974) calls this lifting of constraints 'anti-structure': where individuals experience the 'liberation of human capacities of cognition, affect, volition, creativity, and so on, from the normative constraints' of social status, roles, memberships of groups and social categories (Turner, 1974: 75). Liminal experiences are 'sensorily intense' (Spiegel, 2011: 16); as well as being experienced as liberating, as Turner (1974) indicated, the absence of social structures that we rely on to accomplish things can also provoke feelings of ambivalence, vulnerability and anxiety.

Söderlund and Borg's (2018) recent review of liminality in management and organization studies identifies 'place' as a core theme in what is now a large literature on liminality. They define liminal places as 'spaces created as liminal scenes in which traditional routines, norms and activities are suspended or renegotiated. [...] where activities can unfold in a manner that is unconstrained by conventional norms and traditions' (Söderlund and Borg, 2018: 891). Examples are from Shortt (2015), who shows how moving away from the official front stage in the workplace into marginal, unseen spaces gives workers a pleasing chance to drift and temporarily inhabit 'a more autonomous non-corporate identity' (Shortt, 2015: 653). Daskalaki et al's (2016: 192) study of trans-local work shows how repeated boundary crossing implants multiple sets of norms within the traveller, resulting in an uncomfortable perpetual liminality where 'one is never fully emplaced or fixed, and one is never just "I" but also "an-other"'.

A striking and, for us, highly relevant example of a liminal space comes from Prasad (2014). While researching the Occupied Territories but staying in Israel, he needed to cross the Qalandiya checkpoint twice every day to conduct fieldwork. He thus shared, in part vicariously, the experience of Palestinians crossing the border for their work in Israel. Passing through the checkpoint involved being forced into a narrow, crowded tunnel staffed by armed officers, being subjected to capricious security checks and hearing from Palestinian workers stories of awful violence. For Prasad (2014), the border was a liminal space, one that was outrageous and deeply troubling, but also transformative in the sense that it redefined 'both the ontologies that I was constituted by and the ideologies for which I stood' (Prasad, 2014: 233). In our previous research on how solidarity with refugee women living in Bourj al Barajneh is made, we also recorded a feeling of being lost for words, aghast, when coming face to face with the conditions there, and have changed as a result (Schwabenland and Hirst, 2022). As researchers, we can leave, and our experiences have occasioned lasting, settled changes to our sense of self, but the experiences of Palestinians cannot similarly be resolved. A Palestinian woman in Peteet's (2017) ethnography reports her everyday experience of fearful anticipation of passing through the checkpoint, being in effect 'stripped' of human status. For her, it is a liminal space, but 'there is no celebration of a new status, simply passage from one space to another' (Peteet, 2017: 120).

As the experiences of liminal spaces already mentioned document, liminal spaces entail bodily and sensual experiences, they evoke strong emotions and enable, at some level, a reconstitution of the self. In her theorization of affective ethnography, Gherardi (2019: 749) observes that 'space is a particularly important vehicle for the transmission of affect'. We consider affect next.

Affective practices

In order to use the concept of affect productively to understand how socio-spatial boundaries might be redrawn, we need to be precise about what we mean. We admit that, when first approaching the concept, we spent some time feeling confused: like Cresswell (2013: 230), we found that 'it is not easy to say exactly what the term *affect* refers to'. The field seemed to be littered with circular definitions and a fair bit of bad biology. We found helpful critiques of affect and emotion in geography (Thien, 2005; Pile, 2010; Cresswell, 2013). But the source which has most helped to clarify our ideas is Wetherell's (2012) *Affect and Emotion*, which, as well as doing a ruthless decluttering exercise on what she sees as the confusions in the field, develops the concept of *affective practice* – a term that Wetherell (2012: 23) prefers because practice implies 'ongoingness – patterns in progress', rather than some fixed thing. Wetherell (2012: 159) defines affective practice as: 'A moment of recruitment and often synchronous assembling of multimodal resources, including, most crucially, body states. It is the participation of the emoting body that makes an assemblage an example of affect rather than an example of some other kind of social practice.'

We will unpack four ideas from this definition. First, affect is always relational and intersubjective: it is not located inside the self, but is 'a relation to others, a response to a situation and to the world' (Wetherell, 2012: 24). Second, affective practices are done with and by 'the emoting body', as Wetherell (2012: 159) puts it, but always in conjunction with a wider congregation of human and non-human components of any ontology. While people purposefully bring assemblages together in pursuit of their projects (such as installing suitably equipped kitchens, staff, recipes, ingredients), the concept of assemblage also refers to their happenstance, open-ended, baggy character, as Sohn's (2016: 184) definition, 'heterogeneous and open-ended groupings of material and semiotic elements that do not form a coherent whole', implies. We will use the term assemblage as this is now widely established, but we also note the original French word *agencement* from which it was translated (originally from Deleuze and Guattari, 1987), which directly conveys the activity and vitality of both human and non-human participants in mobilizing and recruiting others.

Third, the question of *how* affect becomes a property of a collective is important for understanding its power, but is a contested one. The 'spread' of particular affects is often described in terms of flows of affect, or as affective atmospheres that we become caught up in, or that are contagious. Metaphors such as flow and contagion are seductive, but they are problematic (Thien, 2005; Pile, 2010; Wetherell, 2012) – they represent affect as if it were a gas or infectious virus that infuses our bodies and leads us to act upon its emotional content, whether we decide to or not. We agree with Wetherell (2012: 141), who argues that affective practices become collective ones through a process of attunement, an ongoing mind–body accomplishment made through 'rapid, implicit and explicit, negotiation processes through which we jointly begin to figure the affective moment we are in, and what should happen next'. Wetherell (2012: 148) further suggests that 'social identity and the social practices associated with that identity are clues to the limits of affective communication'.

Fourth, affective practice involves both repetition and improvisation. It is always 'turned on and simmering', but comes in and out of focus. Affective patterns can 'last, and be reworked, over many hundreds of years', forming 'affective ruts' (Wetherell, 2012: 12–13). Indeed, Thien (2005) detects a whiff of misogyny in accounts of affect that privilege spectacle and intensity in contrast with feminized investigations of domesticated emotion.

This is relevant to our case, a situation marked by long-standing ruts and domestic spaces.

Methodology

The data we present here come from a larger study of women's social enterprises in Palestinian refugee camps. Christina has carried out fieldwork in the two camps we discuss, Bourj al Barajneh and Aida Camp. During her fieldwork visits Christina interviewed women working in non-governmental organizations (NGOs) and social enterprises, carried out more informal conversations with camp residents, visited the kitchen and garden in Bourj al Barajneh, stayed overnight in an NGO-run guesthouse in Aida Camp and kept a diary recording her impressions of her experiences. Access was arranged through previous contacts. Although the UK Foreign Office advises travellers not to visit the camps unless necessary (www.gov.uk/foreign-travel-advice), Christina experienced camp residents to be friendly and welcoming, and generous with their time and hospitality. And although we have refrained from naming individuals, participants were very eager that the names of their organizations be included in any publications, as they want their situation to be better known and understood.

Our dataset consists of interview transcripts, diary notes and photographs alongside secondary data from reports, websites and Facebook entries. Conversations took place in English, with some translation provided, as needed, from NGO workers. Analysis was conducted by Christina and Alison jointly, working as an 'insider–outsider pair' (Lingard et al, 2007), using the approach to identifying key themes advocated by Braun and Clarke (2013) and involving continuous interplay between theoretical resources and empirical materials (Van Maanen et al, 2007). This was used with both the primary, 'richer' data assembled in the camps and the secondary data, including online data. Coding was attentive to both conversational and observational evidence for the ways in which people talked about and used the two different sites.

Findings

Soufra's kitchen: liminality and transformations of self

Our first example is the Soufra kitchen (Figure 8.2), which is housed within the Women's Program Association inside the boundaries of Bourj al Barajneh camp. The kitchen is well equipped, with industrial-sized ovens and grills, but its design uses units and fixtures that seem 'homely', more in keeping with a domestic kitchen than a catering business. Inside this kitchen the women combine fresh ingredients, recipes and *techne*, the knowledge of

Figure 8.2: Inside the Soufra kitchen

Source: Christina Schwabenland

how to cook, and of necessary professional and hygiene standards; also the norms and patterns of working together.

Christina visited the kitchen on two occasions. Her first impressions included the sight and smells of appetizing food (confirmed when she was given a freshly cooked pastry to eat), and an atmosphere of busyness and quiet concentration, interspersed with occasional comments and laughter. The women worked together seamlessly, moving around the kitchen confidently and competently; they did not bump into each other or obstruct each other's movements, and they inhabited the space with poise and self-assurance. At one point the electricity went off and nobody panicked, even though there were pastries in the oven, and sure enough, a minute or so later, the generator kicked in. A set of affective practices have thus emerged where the women's shared history of working together, including coping with such difficulties, has enabled them to form harmonious, 'collective ways of figuring situations' (Wetherell, 2012: 43), with related bodily actions and responses.

The women transform these ingredients into appetizing, healthy dishes based on traditional Palestinian recipes that the women's mothers and grandmothers might have used and brought with them from Palestine into the camp. Preparing these dishes thus reproduces affective practices of family cooking and eating that have endured over generations, long before the displacement. However, the women are not producing this food for their families, but to sell – sometimes the customers may be family and friends within the camp, but they also take food out via a food truck, to sell in markets and at catering events.

Our suggestion that the kitchen is a liminal space comes partly from its design: it is both domestic kitchen and industrial workplace. For it to be a liminal space, however, it must be a locus where transformations occur (beyond the actual preparation of food). The Director explained how such transformations became possible:

> '[I]nside the camp, women, they are not very empowered. And men, in general, they don't like women to work. They say you are a woman you have to stay at home, with the kids, cleaning […] In Soufra they are still in the same environment as the home, they are cooking, they are not going out to work […] so they are inside the camp [and] when the husband wants his wife, he can just pass by and check that she is inside the kitchen, and she is working, and the kind of work is familiar to them, because women know how to cook.'

Here we see the women juggling multiple roles, skillfully evoking them to achieve different ends: for their husbands, they still are wives and mothers, roles which make it difficult for them to work outside the home (Peteet, 2005); for each other and the managers of Soufra, they are professional chefs and businesswomen. Thus, it is the mutability of the Soufra kitchen, as both an extension of the home and also the means by which the Soufra cooks can leave the home and generate income as businesswomen that makes it possible for women to combine these roles. The presence of the women in the kitchen is reassuring and familiar to their husbands; it does not disrupt the long-established affective routines, "because women know how to cook".

And yet, working in the kitchen also changes the ways the women see themselves and are seen by families and colleagues. For example, one woman was cited as saying "now I become to address my clothes, before, I was all the time in my pyjamas". Her husband and family also 'see' her differently: "the husband can see that she is going out of the house for two or three hours, she is coming back happy and motivated, and she […] is bringing income. And this income might be for the home, for herself, for her kids, whatever" (interview with the Assistant Manager). These transformations may be read as radical in a culture where women's roles are circumscribed by traditional expectations. But they are achieved in ways that are quiet and measured and provide reassurance to family members through the kitchen's location in the camp and its evocation of domesticity, and to customers through the professional organizing of tasks and the communication of the quality standards that they enact. Here is a mix of things that are consciously designed and intentionally located, recombining with the various meanings that the food and the act of cooking together hold for them.

Rooftop gardens: nurturing activism and reclaiming dignity

Our second examples are two rooftop gardens in both of the camps Christina visited – one on the rooftop of the Lajee Centre at Aida Camp and the other on the rooftop of Women's Program Association in Bourj al Barajneh. Their rooftop locations are a pragmatic response to the lack of land available for cultivation in the camps, and the difficulties in accessing freshly grown produce. These are hydroponic gardens; again a sensible response to a land that often has little rainfall, using '70% less water than traditional methods of raising garden produce' (www.lajee.org).

The Lajee garden (Figure 8.3) grows a wide variety of crops, including 'tomato, parsley, mints, lettuce, strawberry, and onions, all 100% chemical-free and organic […] additionally, the garden offers an opportunity for an older generation knowledgeable in raising crops to work with younger generations' (www.lajee.org), thus connecting people across generations, receiving and handing down life-sustaining and culture-affirming materials and practices. The garden produces cheap, fresh, healthy food, produced by the women who use the Lajee Centre, for their families and friends. The older women instruct the younger gardeners in the uses of the different plants; a video on the Lajee website shows one woman rubbing a leaf to bring out 'the scent of home' (https://lajee.org/?playlist=00595baandvideo=0b8aab1), telling the stories of the garden her mother tended before she had to leave her village.

Aida Camp is located just outside Bethlehem and its entrance is directly opposite the Barrier Wall. A road connects the two and finishes at a gate in the wall which leads to a military base. The wall separates Aida camp from the surrounding hills and villages, although they are still visible from the rooftops. For Peteet (2017: 10), this is 'closure' – the removal of land from Palestinians, which can then be repopulated and relandscaped.

When Christina visited in 2018, the Lajee garden was quite small, with several plant beds (just distinguishable in Figure 8.3) and a small greenhouse

Figure 8.3: View from the Lajee rooftop garden

Source: Christina Schwabenland

that was badly damaged, covered by a plastic tarpaulin When she asked why, the coordinator of the Women's Unit told her that it had been shredded by tear gas, sprayed from armoured vehicles that come out of the entrance to the wall at night and drive through the camp. Christina noted in her research diary feeling a visceral catch in her stomach and a quick intake of breath; 'the casual destruction of these little seedlings; I feel impotent and furious' (diary fieldnotes).

And this was not a one-off occurrence. Although the garden was repaired and extended, and, in an act of 'cruel optimism' (Berlant, 2010), officially launched in 2021, in January 2022 another attack 'destroyed 1,000 seedlings and extensively damaged the garden' (www.lajee.org). A report published in 2018 into the use of tear gas in Aida Camp found that '100% of residents surveyed reported being exposed to tear gas in the past year; over the same period, 84.3% were exposed in the home, 9.4% at work, 10.7% in school' (Haar and Ghannam, 2018: 2).

The Lajee garden is composed through a multiplicity of affective practices – the reclamation of 'land' on the roofs of camps too densely built to provide access to the earth nurturance of the seedlings, the provision of the healthy fresh fruit and vegetables and the intergenerational transmission of traditional horticultural practices. This garden functions as a liminal space; the affective practices of repairing, restoring and replanting it, following these attacks, transform the gardeners into resilient activists,

Figure 8.4: Author in the rooftop garden of the Women's Program Association

Source: Christina Schwabenland

refusing to accept its destruction alongside the ever-present threats of attack (for the camp residents) and provocation (the Israeli army). The women who tend the seedlings are occupying a traditional role: growing food for their families and friends, nurturing the little plants. But they are also nurturing resilience and resistance, and affirming their cultural heritage and, as the older woman featured on the video clip confirms, their right to healthy, life-giving food. The garden has enabled women across the generations to raise awareness of the conditions they are living under and to advocate for their cause.

The garden on top of the Women's Program Association (WPA) in Bourj al Barajneh (Figure 8.4) has (so far) a happier story. Again, developed as a response to the high density of building in the camp (estimates of between 30,000 and 50,000 people living in one square kilometre), the WPA garden has grown very quickly since it was first developed in 2018. It now provides fresh produce for Soufra and to sell within the camp. It also functions as a meeting place. The garden is a site of conviviality, where local women meet and drink coffee, relax and bring their children, alongside the gardeners tending to the beds of plants. The WPA also hosts visitors from the UN and other international donor organizations in the garden, and this allows the women who work there to take on the role of host (as well as recipient).

The rooftop locations of these gardens disrupt the more conventional associations we make about the relationship between a garden and the earth, especially as here, unlike, say, the more wealthy inhabitants of urban rooftop gardens in Manhattan or other densely populated cities, who have more choices open to them, the access to land is constrained or denied to the residents of the camps. By using hydroponics, they are demonstrating resilience and a commitment to a more ecological and future-looking technology. Both are sites of transformation: the destruction of the seedlings alongside the valuing and passing down of ancient skills of husbandry; the reversal of roles from dependent to host.

Conclusion

In our consideration of the kitchen and the rooftop gardens, we have sought to follow Wetherell's (2012: 52) advice to identify 'the affective assemblages operating in important scenes in everyday life along with their social consequences and entailments'. In both cases, the affective assemblages involved have created openings in what were previously constraining boundaries, in ways that have expanded opportunities for women living in the camps. Women who work in the kitchen or garden can be, simultaneously, wife and chef, daughter and gardener, or host to important guests, or social media representative. When the husband walks by the kitchen he looks in and sees his wife; the women in the kitchen see each other as chefs: both identities are present. The spaces are liminal, not in the sense that all everyday norms associated with women's private roles are suspended, or that distinctions between private and public are collapsed, but in the sense that new public roles coexist simultaneously with private roles.

Cultivating vegetables, and cooking and providing food are traditionally categorized as feminized, domestic activities, imbued with affective meanings of care, nurturance and attention to the needs of others. Here they have been reframed and imbued with political significance. They communicate resilience, pride in their heritage and progressive thinking. These improvisations in affective practices have reconfigured the 'affective ruts' (Wetherell, 2012: 13) laid down for many decades, not through a spectacular or sensational change but by small, iterative, gentle, changes that build on but do not directly challenge women's traditional roles. They stir positive emotional responses, but quietly.

The liminal spaces of the kitchen and gardens are material thresholds as well as symbolic ones – they are gateways between the enclosed space of the camp and the outside world. We have noted the positive entailments of the creation of such openings, in terms of the new social identities available to women working in and performing them. But their exposure to the outside also makes the spaces vulnerable, as the repeated tear-gassing of the Lajee garden indicates. As Wetherell (2012) points out, affect isn't straightforwardly contagious: attunement to affective practices is a matter of choice, and our social identities and associated bodily dispositions play a significant role in how we choose. The seedlings in the Lajee garden make us feel like looking after them, but the Israeli tank drivers who tear-gassed the garden did not feel likewise: they felt something, although we don't have the data to say what. So far, the residents of Aida Camp have kept replanting and rebuilding the garden, so that the new affective practices it creates keep being re-established, adding a sense of resilience and determination to the mix. As Highmore (2010: 123) remarks, '[m]uch of what constitutes the day-to-day is irresolvable and desperately incomplete, yet, for all that, most vital'. The garden is not resolved – it has to be remade effortfully, time and again, but is all the more vital for being replanted.

In this chapter, we have celebrated instances of the improvisation of affective practices by women doing extraordinary things, enabled by the assembling of ordinary domestic spaces. As researchers, we are caught up in the circulation of affects: we are positively attuned to their practices of cooking, growing, and expanding their choices as women. We recognize that we too are caught up in a sense of 'cruel optimism' (Berlant, 2010) in the sense of hoping that the positive changes we have seen could themselves become established 'affective ruts' (Wetherell, 2012: 13), practices that it is just normal for women to do as part of their everyday, dignified working lives. Meanwhile though, Palestinian land continues to be settled illegally and the chance of a wider political settlement is diminishing. These other affective ruts are being ploughed ever deeper.

References

Abu Mughli, M. (2020) Education and Palestinian Refugees in Lebanon. *Education, Occupation and Liberation* webinar organized by Friends of Bir Zeit University (FOBZU). Available at: https://fobzu.org/blog/2020/11/05/education-and-palestinian-refugees-in-lebanon-an-interview-with-dr-mai-abu-moghli/ (accessed 17 November 2023).

Berlant, L. (2010) Cruel Optimism. In M. Gregg and G. J. Seigworth (eds) *The Affect Theory Reader.* Durham and London: Duke University Press, pp 93–117.

Boyce, G., Marshall, D. J. and Wilson, J. (2015) Concrete Connections? Articulation, Homology and the Political Geography of Boundary Walls. *Area*, 47(3): 289–295.

Braun, V. and Clarke, V. (2013) *Successful Qualitative Research: A Practical Guide for Beginners.* London: Sage.

Cresswell, T. (2013) *Geographic Thought.* Chichester: Wiley-Blackwell.

Daskalaki, M., Butler, C. L. and Petrovic, J. (2016) Somewhere In-between: Narratives of Place, Identity, and Translocal Work. *Journal of Management Inquiry*, 25(2): 184–198.

Deleuze, G. and Guattari, F. (1987) *A Thousand Plateaus: Capitalism and Schizophrenia*. Minneapolis, MN: University of Minnesota Press.

Gherardi, S. (2019) Theorizing Affective Ethnography for Organization Studies. *Organization*, 26(6): 741–760.

Haar, R. and Ghannam, J. (2018) *No Safe Space: Health Consequences of Tear Gas Exposure amongst Palestinian Refugees*. Berkeley, CA: Human Rights Center, University of California.

Hanafi, S., Chaaban, J. and Seyfert, K. (2012) Social Exclusion of Palestinian Refugees in Lebanon: Reflections on the Mechanisms that Cement Their Persistent Poverty. *Refugee Survey Quarterly*, 31(1): 34–53.

Highmore, B. (2010) Affect, Food, and Social Aesthetics. In: M. Gregg and G. J. Seigworth (eds) *The Affect Theory Reader*. Durham, NC and London: Duke University Press, pp 118–137.

Lingard, L., Schryer, C. F., Spafford, M. M. and Campbell, S. L. (2007) Negotiating the Politics of Identity in an Interdisciplinary Research Team. *Qualitative Research*, 7(4): 501–519.

Massey, D. (2005) *For Space*. London: Sage

Peteet, J. (2005) *Landscape of Hope and Despair*. Philadelphia: University of Pennsylvania Press.

Peteet, J. (2017) *Space and Mobility in Palestine*. Bloomington, IN: Indiana University Press.

Pile, S. (2010) Emotions and Affect in Recent Human Geography. *Transactions of the Institute of British Geographers*, 35(1): 5–20.

Prasad, A. (2014) You Can't Go Home Again: And Other Psychoanalytic Lessons from Crossing a Neo-colonial Border. *Human Relations*, 67(2): 233–257.

Schwabenland, C. and Hirst, A. (2022) Solidarity with Soufra: Dividuality and Joint Action with Palestinian Women Refugees. *Organization*, 29(2): 324–344.

Shortt, H. (2015) Liminality, Space and the Importance of 'Transitory Dwelling Places' at Work. *Human Relations*, 68(4): 633–658.

Söderlund, J. and Borg, E. (2018) Liminality in Management and Organization Studies: Process, Position and Place. *International Journal of Management Reviews*, 20(4): 880–902.

Sohn, C. (2016) Navigating Borders' Multiplicity: The Critical Potential of Assemblage. *Area*, 48(2): 183–189.

Spiegel, A. D. (2011) Categorical Difference Versus Continuum: Rethinking Turner's Liminal-Liminoid Distinction. *Anthropology Southern Africa*, 34(1-2): 11–20. DOI: 10.1080/23323256.2011.11500004.

Thien, D. (2005) After or beyond Feeling? A Consideration of Affect and Emotion in Geography. *Area*, 37(4): 450–454.

Turner, V. (1974) Liminal to Liminoid, in Play, Flow, and Ritual: An Essay in Comparative Symbology. *Rice Institute Pamphlet–Rice University Studies*, 60(3).

van Gennep, A. (1960) *The Rites of Passage*. Chicago: University of Chicago Press.

Van Maanen, J., Sørensen, J. B. and Mitchell, T. R. (2007) The Interplay between Theory and Method. *Academy of Management Review*, 32(4): 1145–1154.

Wetherell, M. (2012) *Affect and Emotion: A New Social Science Understanding*. London: Sage.

9

Placing Fear of Crime: Affect, Gender and Perceptions of Safety

Murray Lee

Introduction

This chapter explores fear of crime as affective, or at least as the outcome of, or response to, *affect*. In doing so it attempts to provide a critical commentary on the way research on perceptions of crime has approached fear of crime reductively, and in many ways sought to collapse complex sociocultural processes into decontextualized (often numeric) research findings. While neither gender, place nor affect has been ignored in the corpus of perceptions of crime research, each has been rendered somewhat subservient to the methods deployed to assess them. Fear of crime has generally been measured, assessed and discussed through the use of large-scale cross-sectional surveys. While, when designed with nuance, such surveys can produce data offering general insights into perceptions and subsequent behaviours down to neighbourhood level, they provide little insight into how individuals' affective engagement, encounters and resonances with places and spaces serve to non-consciously influence negative (or indeed positive) perceptions of safety. The proxies for affect in such surveys tend to confuse it with emotion – and, even more reductively, deal with it simply as a component of fear (which to an extent it is, but it is of course also much more than this).

The chapter is structured as follows. In the first section, I offer an overview of some of the key developments in research on fear of crime, particularly as it relates to gender. The second section explores the components of fear of crime further, and particularly focuses on the concept of affect and how this has been deployed within the fear of crime literature. The third section provides a more compressive and sophisticated overview of affect, and attempts to explain that traditional fear of crime research is misguided and partial in its deployment of affect. The final section looks at place and fear of crime and overlays some of the discussion of affect in order to provide a sensorial and intersubjective perspective on fear of crime, place, gender and affect. Each section begins by drawing on a range of research-respondent quotes from empirical projects conducted in Sydney, Australia. I have chosen not to specifically interrogate or analyse these but to leave them as fragments that speak to the theoretical discussion. In short, the chapter assesses how the environments we engage and interact with provide non-conscious cues and codes for individuals and dividuals that serve to (re)produce gendered hierarchies of security and safety.

Fear of crime and gender

'I absolutely always wear shoes that I can run in. I try to wear things that would be easy to move quickly in, so trousers, not tight jeans or short skirts. I think sometimes, if it was legal to carry items for self-defence that would feel better.' (R33: female, 19)

Understanding, and particularly attempting to measure, fear of crime has had a long and chequered history, largely due to the design and implementation of victim surveys (Hale, 1996; Ditton and Farrall, 2000; Lee, 2001; 2007). While a well-calibrated, large-scale population sample survey can give us some general sense of the levels of crime concern or perceptions of risk, even in regard to specific offences such as burglary, assault or robbery, they tell us little about the mechanics of how such concerns manifest subjectively and intersubjectively *in situ*. That is, they create a number of imagined scenarios and ask respondents to place themselves in these, ascertaining the level of worry experienced in each. In doing so, context is lost, and while respondents no doubt project images of their specific temporal, geographic and spatial contexts of worry onto these more generalized survey items, these are imagined projections. They reflect not some kind of reified phenomenological experience but, rather, layers of cultural meaning and structured social expectations. It is no wonder that researchers have been able to prove that they often both overstate levels of fear and yet understate the complex ways in which fear of crime manifests.[1] That said, in many ways surveys have become an easy target of criticism and their limitations have been well rehearsed in the criminological literature, including by myself (Ditton and Farrall, 2000; Hollway and Jefferson, 2012b; Lee, 2007; 2009). By their very nature, victim surveys remove fear of crime from places and spaces and reduce such spaces and places to standard items asked by the researcher. In short, they reduce complex human experiences to numbers. Nonetheless, in broad terms they rightly highlight gender as a key variable in experiences of worry, even as they might misinterpret and misplace the causes and reasons for this worry.

Indeed, one almost universal finding in fear of crime scholarship is that women report higher levels of fear than men (Skogan and Maxfield, 1981; Stafford and Galle, 1984; Warr, 1984; Ferraro, 1995; Fisher and May, 2009; Franklin and Franklin, 2009). While other social groups likewise report higher levels of fear – the elderly, for example (Greve et al, 2018) – understanding the gendered dimensions of worry has largely driven the entire research agenda. It has been termed the fear of crime paradox – that those least likely to be victims, women and the elderly in particular – appear to be the most fearful. As Maxfield (1984: 47) noted almost 40 years ago, 'the differences in fear by sex and age are so substantial that something other than direct victimization experience must be involved'. In explaining this something else, researchers in the 1980s and 1990s largely postulated two competing theses, which we might call the vulnerability thesis and the hidden victimization thesis.

More traditional empirical scholars noted that the 'spectre of sexual assault' as a master offence hung over women, increasing their gendered sense of unease and sensitizing them to worry about crime even though their risk of victimization was lower (Warr, 1987; Killias, 1990; Ferrarro, 1995). This is the vulnerability thesis. Surveys seemed to tell us that perceived risk and perceived vulnerability to victimization were strongly related. There are physical, social and situational dimensions of vulnerability related to three aspects of threat: 'exposure to non-negligible risk', 'loss of control' and 'seriousness of consequences' (Killias, 1990). This suggests that perceived risk (perceived likelihood, perceived consequence and perceived control) explains to some extent the association between socio-demographic markers of vulnerability (such as gender and age) and worry about crime. For example, Jackson (2009) found that women tended to worry more frequently than men, partly because they felt less able to physically defend themselves; they had higher perceived negative consequences; they had lower perceived control over situations in which they feel vulnerable; and they saw the likelihood of victimization as greater for themselves and for their social group.

On the other hand, feminist and critical scholars suggested the 'actual levels of victimization' for women were far higher than captured in the official statistics (Stanko, 1990). They pointed to the under-reporting of sexual assault and domestic violence and, importantly, to the 'everyday violence' or low-level street, workplace, leisure-time and domestic harassment that women face on a day-to-day basis (Young 1987; Stanko 1990). Once this was accounted for, they argued, women's fear was not some irrational or even understandable flow-on from the spectre of sexual assault, but the result of ongoing experiences of very real victimization (Walklate, 2011). Quite simply, women's victimization was ongoing and largely hidden.

However, by the early 2000s a third identifiable position also began to emerge. This might be called the 'doing gender thesis'. This position largely accepts the gendered gap as found in surveys but places it in a structural social and cultural context. For example, Sutton and Farrall (2005) were able to show that men and women answered survey questions differently in order to meet socially desirable norms – in particular, men underplayed their fears through 'deceptive responding'. While Sutton and Farrall (2005) suggest that the gender gap largely disappears once deceptive reporting is accounted for, others also note that both culture and gender identity are dynamic. Cops and Pleyiser (2011), for example, found that while not completely explaining the gendered gap in fear of crime, the extent to which a respondent reported being more or less feminine or masculine was likely to influence their reported levels of worry.

One commonality in all of this (largely) survey-based research is a search for some kind of objective rationality (or unreason) for fear of crime generally, and the fear gender gap specifically. I have argued elsewhere that the construction of women's fear as somehow excessive or irrational in early fear of crime research was conceptually related to the (usually) unconscious positioning of men's fear as reasonable or, more to the point, natural – against which women's fear was measured (Lee, 2007). Thus, this excessive fear could be seen as the problem to be assessed and managed, resulting in years of fear of crime research focused on its gendered nature. Indeed, this paradox largely drove the fear of crime research agenda as researchers sought to decode women's fear. But what if fear of crime is neither rational nor something respondents are even conscious of? What if at least some of its antecedents are simply unamenable to survey research?

In the late 1990s and early 2000s qualitative research methods also developed around the fear of crime research agenda. Tulloch et al (1998a, 1998b; Lupton and Tulloch, 1999) questioned any direct link between 'fear' and 'crime' and suggested, drawing on anthropological work on risk and dangerousness (Douglas, 2003), that a range of social, economic, aesthetic and existential biographical variables could influence one's position in regard to fear of crime. They note:

> An important development in understanding fear of crime is the positioning of people as reflexive subjects who experience and respond to crime via communal, aesthetic and shared symbolic meanings. We use the term 'reflexivity' here to encompass not simply a process of rationalist self-monitoring through cognitive or normative categories. ... Aesthetic or hermeneutic reflexivity is rooted in background assumptions and unarticulated practices and in intuition, feeling, emotion and the spiritual. This type of reflexivity involves the processing of signs and symbols rather than

simply 'information'. Aesthetic reflexivity relies upon an individual's membership of a community, moral and culturally learned and shared assumptions, preferences and categories. (Lupton and Tulloch, 1999: 512)

Hollway and Jefferson (1997a, 1997b, 2012a), and later Gadd and Jefferson (2009), also investigated the fear of crime qualitatively, deploying psycho-social psychoanalysis frameworks. They suggested that emotional responses to anxiety are intertwined through complex biographies of our individual lives. Founded on post-structuralist theory, they argued that for some individuals the intersubjective defending against anxiety can be central to understanding fear. As Gadd and Jefferson (2009) put it:

In essence then, what we are arguing is that subject positions are negotiated in relation to the individual's biography and attendant anxieties, the discursive fields available to the individual (often constrained by their class, ethnicity and gender), and intersubjectively through the responses to others. Whether someone invests in the position of the fearful subject preoccupied with the ever-growing threat of victimization depends, in part ... on how available that position is to him or her. (Gadd and Jefferson, 2009: 139)

Both approaches developed a more nuanced and contextual reading of crime fear that sees it embedded in cultural norms, physical and material places and individual experiences, while also noting its structural preconditions. More recently, feminist fear of crime research has continued to produce important qualitative insights (Pain and Smith, 2008; Walklate, 2011, 2018; Fanghanel, 2018; Vanderveen, 2018; Fileborn, 2022). Fanghanel (2018), for example, theorizes that rape myths, particularly that of the stranger serial rapist, serve to recreate gendered spatial hierarchies and that this is both structural and enabled through the violence of the state.

The myth of the stranger-rapist who lies in wait to ambush unsuspecting women in dark alleys or from behind a bush is an important component of rape myths more broadly ... [it] continues to construct public space and connect with other machines which produce fear of crime. We cannot accept this because we need to challenge the 'taken-for-grantedness' of rape culture that enables it to thrive as part of State apparatus. (Fanghanel, 2018: 425)

Fileborn (2019), while not seeking to downplay gendered differences in relation to street harassment, also draws our attention to the intersectionality of experiences. She aims to de-essentialize and de-center our understandings of street harassment as a purely gender-based phenomenon (though this is certainly not to say that gender is irrelevant or unimportant – rather, [...] gender as also shaped in and through factors such as race, class, and sexuality), [... and] differentially located, contextual, and fluid experiences. (Fileborn, 2019: 223)

In short, qualitative research, and feminist research in particular, has revealed something of the complexity of the 'experiences' and 'expressions' (Farrall et al, 2009) that shape fear of crime and its relationship to gender in place. Before I explore this further, some explanation is required as to what researchers have generally seen as the components of fear of crime. This explanation will also help to introduce the notion of *affect* in fear of crime research.

Worry about crime and the ABC of fear

'I suffer from anxiety and I am naturally really cautious; being sexually assaulted is my biggest fear in life, so I'm constantly aware of that. Even if it's in the daytime I'm constantly looking out, but I think that's my own fear. Even on buses and trains I just am constantly – if it's just me and another person on the train and stuff, I'm just a bit more aware. But I feel like that's my own anxiety, I don't think that's something everyone would feel. I just get a little bit worked up and afraid sometimes.' (R46: female, 20)

Berenbaum (2010: 963) refers to worry as repetitive and anxiety-producing thoughts that have three characteristics: '(1) the repetitive thoughts concern an uncertain future outcome; (2) the uncertain outcome about which the person is thinking is considered undesirable; and (3) the subjective experience of having such thoughts is unpleasant'. In relation to fear or perceptions of crime, this worry is generally captured by the ABC of fear of crime; *affect, behaviour, cognition* (Gabriel and Greve, 2003; Farrall et al, 2009; Jackson, 2009, 2011. These components of fear are generally explained as follows, as I have noted elsewhere (Lee et al, 2020).

Let's start with *behaviour*. There are four main behavioural responses to the perceived threat of victimization: avoidance behaviour, protective behaviour, behavioural and lifestyle adjustments and participation in relevant collective activities (Miethe, 1995: 21–26). Survey questions that seek to understand behavioural responses tend to focus on things like avoidance of public transport or public places at night, carrying a weapon or installing an alarm, moving to a quiet suburb or joining a community group. Research has often concluded that many such behaviours are likely to be gendered.

The *cognitive* aspects of fear of crime go to assessments of the personal crime risk or harm. Criminological work has shown that the perception of the likelihood of victimization is strongly correlated with expressed levels of fear, worry and anxiety about the event occurring (Ferraro, 1995; Farrall et al, 2009). Also important is people's sense of the seriousness of the consequences if they were to fall victim, as well as their sense of control over its occurrence (Jackson, 2009). This work suggests that worry and anxiety are often sustained by a sense of risk, threat and vulnerability, as discussed earlier (Killias, 1990; Hale, 1996; Gabriel and Greve, 2003; Farrall et al, 2009; Jackson, 2009; 2011).

Finally, to *affect*. Cultural scholars and social scientists have become increasingly interested in the concept of affect, and this has not been ignored by fear of crime researchers. The affective components of 'fear of crime' are emotive responses to the perceived possibility of victimization, including (but not limited to) physical responses to immediate threat ('fear'), repetitive thoughts about future uncertain harm ('worry') and a more widespread but diffuse low-level emotion about uncertain future events that is separate from concrete feelings of imminent danger ('anxiety'). As noted, criminologists have for a long time realized these as difficult emotions to measure (Hale, 1996). For example, Warr (2000) argues that survey questions like 'how afraid are you about being burgled?', which one might on the face of it assume are capturing people's physical fear responses to the threat of burglary, capture instead (or as well) a more future-oriented general anxiety. After all, the research participants answering this question are unlikely to be physically afraid of being burgled right at that point, at the time of the interview.

Affect and fear of crime

'I think that feeling unsafe and feeling on edge has become the new normal. That's what our expectation is. As much as this place is home, it's a troubled home and we need to be on our guards.' (R7: female, 48)

So, while fear of crime researchers have certainly explored affect, quantitative research has generally been more interested in this in relation to simply emotive responses to worry. As discussed earlier, the rather clinical enumeration of fear (Lee, 2009) via survey instruments tends to reduce the messiness of affect's often telling us little about the subjective and intersubjective complexity of how these feelings arise and manifest in place. To explore this further, there is a need to expand our definition of affect. As Hemmings (2005: 550) suggests, 'affect broadly refers to states of being, rather than to their manifestation or interpretation as emotions'.

In fear of crime research, the focus has been on the latter, rather than the former. She goes on to suggest that these states of being also go beyond the assessment of 'drives' as per psychoanalysis. This also suggests that we need to expand the parameters of some of the qualitative research discussed earlier. Rather, 'Unlike drives, affects can be transferred to a range of objects in order to be satisfied (love may have many objects, for example)' (Hemmings, 2005: 550). In terms of fear of crime, then, affect has implications in relation to not just the way in which worry is felt (emotionally) but *how* it manifests and is projected well beyond the rather deterministic objects of the victim survey model. Moreover, emotion more generally as an organizing concept has been criticized by feminist theorists (Thien, 2005) for its binary relationship to reason. It follows, then, that seeing fear as an emotion immediately genders it through a range of binary historical associations that construct men as rational and women as emotional (also see Lee, 2007).

I am not, of course, suggesting that worry about crime is not gendered, only that the notion of affect as emotion has contributed to over-deterministic models of fear, and a blindness or narrowness to *how* affect functions. As Thien (2005: 451) has noted, 'the emerging geographies of affect continue to reach for the "how" of affect'. So perhaps using the concept of affect, along with place and temporality, can help to provide this more thorough account, and it could shed more light on the gendered dynamics of fear of crime.

Massumi (1993) has explored the concept of affective fear not as emotion but as producing particular emotional responses in subjects. Emotions, he says, are not personal feeling. Rather, they are social and socially produced:

> Emotions and the character types they define are the specific social content of the fear affect as the contemporary human equation. They are derivatives of that equation: secondary expressions (in the mathematical sense) of capitalist powers of existence. Character is the derivative of a power equation. It is power determined, as presence-effect. Emotional makeup is the face power turns toward the predictably unbalanced, saleably empty content of an individual life. (Massumi, 1993: 25)

Thus, an affect is a non-conscious experience of some kind of intensity; an unformed, unstructured potential with the power to guide thought and action (Shouse, 2005). Affect is abstract and cannot be fully realized in language because it is outside of consciousness (Massumi, 2002; Shouse, 2005). Tomkins (1995) likens the affect mechanism to the pain mechanism. Like the pain mechanism experienced in our bodies (that tells us we must do something to protect ourselves and prevent further harm), the affect mechanism amplifies our awareness, activating a response such that we are forced to be concerned, and concerned immediately.

All this significantly expands our notion of *affect* from simply the emotional responses recorded in a survey to an analysis of the emotional, spatial and political (to the pre-conscious, pre-emotional). Yes, fear of crime is about how we feel, our experiences and expressions (Farrall et al, 2009), but these are only the individual and social responses to affect. It should also be about how these feelings and emotions are derived through what we see (or don't see), hear, feel, smell, taste; through histories – social and individual; through intersubjective and subject experiences; through situational cues; and through differential relationships to and of power. As Wrenn (2014: 338) has said of fear more generally, 'it must be studied as a process that develops under its own inertia, feeding off its antecedent and instinctual past, as well as a phenomenon that is shaped by and in turn shapes its institutional setting'.

The problem, then, is that while quantitative fear of crime research can tell us how frequently or intensely (on a quantitative level), respondents express or report experiencing worry (and it can even tell us something about the behavioural responses and evaluations of risk which overlap with these emotional responses), it gives us little insight into the sensorial, the imaginative and the aestheticized (Young, 2014). It accounts for the emotional responses to affect but does not explore its pre-conscious conditions. We can explain this by returning to the way in which rape culture operates in Fanghanel's (2018) model discussed earlier. The affects associated with rape culture are 'pre-personal' and non-conscious (Deleuze and Guattari, 1987), that is, worry about rape is affective – measuring its emotional response tends to problematize the feelings of the respondent rather than a set of power relations that allow its affect to circulate. Feelings of vulnerability may also be a response to rape culture, and here 'pre-personal' affect is (inter)subjectively processed in relation to cognitive evaluations of one's physical capacities, but also in relation to the imaginary of an assault, themselves the result of crime narratives, stories, media reports and personal experiences. A process that develops under its own inertia, to paraphrase Wren (2014), cited earlier.

Place and fear of crime

> 'My area is quite disorderly. We don't go out at night. If we do, we come home in a taxi, quickly get inside my place. ... Overall, the public order is still not good. Scary. I am still scared.' (R67: Cantonese speaker, female)

While surveys take fear of crime out of place, there are qualitative methods that can be more attentive to location and yield differing data (for example,

Vanderveen, 2018). Many researchers have pointed to the fact that fear of crime is situational (Fattah and Sacco, 1989), and that it is influenced by the specifics of locations and the interactions with others in those locations (Pain, 2000; Rengifo and Bolton, 2012). It can be transitory, momentary, or it can linger once one leaves a location. We that cues of neighbourhood disorder can impact on fear of crime (Sampson and Raudenbush, 1999), with particular sensory cues leading to increased worry, such as graffiti, rubbish, unlit areas, unpleasant smells such as sewage or drainage (Inness, 2004). Indeed, such notions can be traced back to Wilson and Kelling's (1982) broken windows thesis and Clark's (1983) situational crime prevention. Moreover, the specific features of the physical environment, such as the capacity for escape or to take refuge, are also said to have an impact (Fisher and Nasar, 1995). Indeed, Jackson (2009: 382) has noted the need for fear of crime research to understand more about how 'dispositional perceptions of risk ... interact with situational risk appraisals (the perceptions that relate to a specific situation to produce momentary worries about falling victim)'.

In many ways such a call aligns with a recent provocation by McClanahan and South (2020) for the development of a 'sensory criminology'. McClanahan and South (2020) build on Keith Hayward's (2012: 442) earlier call to research 'phenomenological place over abstract space in an attempt to take seriously the cultural and structural relationships that contribute to crime and disorder or, for that matter, community safety and stability'. While cultural criminology already has a sophisticated rendering of space, where it is 'understood almost as if it were a living thing, a multi-layered congress of cultural, political and spatial dynamics' (Hayward, 2012: 443), Hayward suggests that cultural geography might provide understandings of how 'landscapes function as systems of social reproduction' (2012: 449). While spaces and places reproduce culture, culture also reproduces place (de Certeau, 1984). Thus, Hayward articulates 'five spaces of cultural criminology' (2012): more-than-representational spaces, parafunctional spaces, container spaces, virtual/networked spaces and acoustic spaces. Among these, 'more-than-representational space' operates symbolically at the level of culture. Such spaces take on meaning and circulate that meaning through structured power relations and intersubjective engagement. Putting aside engagement with Hayward's other categories, for the purposes of brevity, spaces that are more than representational speak to the concept of affect. That is, they contain and circulate affective capacities beyond their material and physical attributes. An empty, poorly lit lane in and of itself is not of concern unless it affectively evokes cultural, aesthetic and sensory cues. Similarly, Young (1996: 15) introduced the notion of an imaginary of crime, 'the process by which we make images of crime', and how such images of crime make us. Using the split between I/eye, she examines the relationship between seeing or not seeing (the eye) and the way subject formation (I) takes place only through the construction of others.

Taking these sensorial and aestheticized tropes further, McClanahan and South (2020) suggest the need to also map sensory spaces (as opposed to physical and material spaces). They argue that

> we can think of the five *interior* sensorial spaces as mapping over the five *external* (though also affective) spaces described by Hayward (emphasis in original). Transposing internal and embodied sensorial spaces (sight, sound, taste, smell and touch) over and across the external spaces elaborated by Hayward offers a new layer to the map of cultural criminological thought, new and intensely affective sites of analysis that we feel strengthen cultural criminology's already well demonstrated commitment to creative ways of thinking about crime, harm, control and power. (McClanahan and South, 2020: 6)

This has important implications for understanding fear of crime and affect in place. Spaces and places are not thus inert, nor are they simply aesthetic physical environments upon which we can read of the cues of fear of crime. Rather, they are affective. And the ways in which these affects influence our individual and collective perceptions about crime are complex and messy. Yes, they are to an extent structured, and gender identity, age, sex, ethnicity and disability are all likely to influence how our bodies experience these affects. But these are filtered through our individual biographies as well as the sensorial cues associated with these spaces and places.

Conclusion

> 'I think some of these scenarios, too, it really depends on your individual perception. Just the relatively minor incident that [other respondent] had the other day, he will shrug it off and go, just a loony, a bit of an idiot, but someone else, that would be crippling for them, they'd be a bundle of nerves, they wouldn't come out of their house for weeks, what's happened to society, everything I'm reading in the newspapers is right. Perception is different.' (R16: male, elderly)

To some extent various scenarios used in survey research also tell us something about the situational triggers for emotional responses to the threat of crime. Moreover, demographic data and even variables such as media consumption can give us further clues as to what groups are most likely to experience such negative emotional responses. However, I have argued here that while all of this might be of some use, fear of crime research has generally failed to theorize its object of inquiry in a way that interrogates its complex antecedents. Indeed, such research has generally explored emotional responses because these antecedents are too elusive. However, with recent theoretical developments the concept of affect could provide some of the conceptual clarity that fear of crime research so obviously needs. The challenge is to move away from the neat survey model and to embrace complexity. This is never easy for a discipline like criminology that seeks to balance theoretical insights with policy relevance.

Note

[1] A more sophisticated survey can also measure the intensity and frequency of worry; how often people remember experiencing worry, and how intense that worry was the last time they experienced it. Farrall et al (2009) found that a substantial proportion of respondents who say they were 'very' or 'fairly' worried also report that they had *not worried once* over the past 12 months. Indeed, a good proportion of those individuals who reported some overall intensity of worry could not recall a single instance of when their emotions surfaced.

References

Berenbaum, H. (2010) An Initiation–Termination Two-phase Model of Worrying. *Clinical Psychology Review*, 30(8): 962–975. DOI: 10.1016/j.cpr.2010.06.011.

Certeau, M. de (1984) *The Practice of Everyday Life*, Berkeley: University of California Press.

Clarke, R. V. (1983) Situational Crime Prevention: Its Theoretical Basis and Practical Scope. *Crime and Justice*, 225–256.

Cops, D. and Pleysier, S. (2011) The Doing Gender in Fear of Crime: the Impact of Gender Identity on Reported Levels of Fear of Crime in Adolescents and Young Adults. *British Journal of Criminology*, 51(1): 58–74.

Deleuze, G. and Guattari, F. (1987) *A Thousand Plateaus – Capitalism and Schizophrenia*, .Minneapolis and London: University of Minnesota Press.

Ditton, J. and Farrall, S. (2000) *The Fear of Crime*, London: Routledge.

Douglas, M. (2003) *Purity and Danger: An Analysis of Concepts of Pollution and Taboo*. London: Routledge.

Fanghanel, A. (2018) How to Break a Rape Culture: Gendered Fear of Crime and the Myth of the Stranger-Rapist. In: M. Lee and G. Mythen (eds) *Routledge International Handbook on the Fear of Crime*, Abingdon: Routledge.

Farrall, S., Jackson, J. and Gray, E. (2009) *Social Order and the Fear of Crime in Contemporary Times*. Clarendon Studies in Criminology. Oxford: Oxford University Press.

Fattah, E. A. and Sacco, V. F. (1989) *Crime and Victimisation of the Elderly*. New York: Springer-Verlag.

Ferraro, K. F. (1995) *Fear of Crime. Interpreting Victimization Risk*. Albany: State University of New York Press.

Fileborn, B. (2019) Naming the Unspeakable Harm of Street Harassment: A Survey-Based Examination of Disclosure Practices. *Violence Against Women*, 25(2): 223–248.

Fileborn, B. (2022) Mapping Activist Responses and Policy Advocacy for Street Harassment: Current Practice and Future Directions. *European Journal on Criminal Policy and Research*, 28: 97–116. DOI: 10.1007/s10610-021-09479-2.

Fisher, B. S. and May, D. (2009) College Students' Crime-related Fears on Campus: Are Fear-provoking Cues Gendered? *Journal of Contemporary Criminal Justice*, 25: 300–321.

Fisher, B. and Nasar, J. J. (1995) Fear Spots in Relation to Microlevel Physical Cues: Exploring the Overlooked. *Journal of Research in Crime and Delinquency*, 32(2): 214–239.

Franklin, C. A. and Franklin, T. W. (2009) Predicting Fear of Crime: Considering Differences across Gender. *Feminist Criminology*, 4: 83–106.

Gabriel, U. and Greve, W. (2003) The Psychology of Fear of Crime: Conceptual and Methodological Perspectives. *British Journal of Criminology*, 43: 600–614.

Gadd, D. and Jefferson, T, (2009) Anxiety, Defensiveness and the Fear of Crime. In: M. Lee and S. Farrall (eds) *Fear of Crime: Critical Voices in and Age of Anxiety*. Oxon: Routledge.

Greve, W., Leipold, B. and Kappesand, C. (2018) Fear of Crime in Old Age: A Sample Case of Resilience? *The Journals of Gerontology: Series B*, 73(7): 1224–1232. https://doi.org/10.1093/geronb/gbw169

Hale, C. (1996) Fear of Crime: A Review of the Literature. *International Review of Victimology*, 4(2): 79–150.

Hayward, K. J. (2012) Five Spaces of Cultural Criminology, *The British Journal of Criminology*, 52(3): 441–462.

Hemmings, C. (2005) Invoking Affect: Cultural Theory and the Ontological Turn. *Cultural Studies*, 19(5): 548–567.

Hollway, W. and Jefferson, T. (1997a) The Risk Society in an Age of Anxiety: Situating Fear of Crime. *The British Journal of Sociology,* 48(2): 255–266.

Hollway, W. and Jefferson, T. (1997b) Eliciting Narrative Through the In-Depth Interview. *Qualitative Inquiry,* 3(1): 53–70.

Hollway, W. and Jefferson, T. (2012a) *Doing Qualitative Research Differently: A Psychosocial Approach*. London: Sage.

Hollway, W. and Jefferson, T. (2012b) The Role of Anxiety in Fear of Crime. In: T. Hope and R. Sparks (eds) *Crime, Risk and Insecurity*. London: Routledge, pp 31–49.

Innes, M. (2004) Signal Crimes and Signal Disorders: Notes on Deviance as Communicative Action. *The British Journal of Sociology*, 55(3): 335–355.

Jackson, J. (2009). A Psychological Perspective on Vulnerability in the Fear of Crime. *Psychology, Crime and Law*, 15(4): 365–390.

Jackson, J. (2011) Revisiting Risk Sensitivity in the Fear of Crime. *Journal of Research in Crime and Delinquency*, 48(4): 513–537.

Killias, M. (1990) Vulnerability: Towards a Better Understanding of a Key Variable in the Genesis of Fear of Crime. *Violence and Victims*, 5(2): 97–108.

Lee, M. (2001). The Genesis of 'Fear of Crime'. *Theoretical Criminology*, 5(4): 467–485. https://doi.org/10.1177/1362480601005004004

Lee, M. (2007) *Inventing Fear of Crime: Criminology and the Politics of Anxiety*. Cullompton: Willan.

Lee, M. (2009) The Enumeration of Anxiety: Power, Knowledge and the Fear of Crime. In: M. Lee and S. Farrall (eds) *Fear of Crime: Critical Voices in an Age of Anxiety*. New York: Routledge, pp 32–44.

Lee, M., Jackson, J. and Ellis, J. R. (2020) Functional and Dysfunctional Fear of Crime in Inner Sydney: Findings from the Quantitative Component of a Mixed-methods Study. *Australian and New Zealand Journal of Criminology*, 53(3): 311–332.

Lupton, D. and Tulloch, J. (1999) Theorizing Fear of Crime: Beyond the Rational/Irrational Opposition. *The British Journal of Sociology*, 50(3): 507–523.

Massumi, B. (1993) *The Politics of Everyday Fear*. Minneapolis: University of Minnesota Press.

Maxfield, M.G. (1984) *Fear of Crime in England and Wales*. London: HMSO.

McClanahan, B. and South, N. (2020) All Knowledge Begins with the Senses: Towards a Sensory Criminology. *The British Journal of Criminology*, 60(1): 3–23.

Miethe, T. (1995) Fear and Withdrawal from Urban Life. *Annals of the American Academy of Political and Social Sciences*, 539: 14–27.

Pain, R. (2000) Place, Social Relations and the Fear of Crime: A Review. *Progress in Human Geography*, 24(3): 365–387.

Pain R, and Smith S. (2008) *Fear: Critical Geopolitics and Everyday Life*. London: Ashgate.

Rengifo, A. F. and Bolton, A. (2012) Routine Activities and Fear of Crime: Specifying Individual-Level Mechanisms. *European Journal of Criminology*, 9(2): 99–119.

Sampson, R. J. and Raudenbush, S. W. (1999) Systematic Social Observation of Public Spaces: A New Look at Disorder in Urban Neighborhoods. *American Journal of Sociology*, 105(3): 603–51. DOI: 10.1086/210356

Shouse, E. (2005) Feeling, Emotion, Affect. *Media/Culture Journal*, 8(6). https://doi.org/10.5204/mcj.2443

Skogan, W. G. and Maxfield, M. G. (1981) *Coping with Crime: Individual and Neighborhood Reactions*. Beverly Hills, CA: SAGE.

Stafford, M. and Galle, O. (1984) Victimization Rates, Exposure to Risk, and Fear of Crime. *Criminology*, 22: 173–185.

Stanko, E. (1990) *Everyday Violence: Women's and Men's Experience of Personal Danger*. London: Pandora Press.

Sutton, R. M., and Farrall, S. (2005) Gender, Socially Desirable Responding and the Fear of Crime: Are Women Really More Anxious About Crime? *British Journal of Criminology*, 45(2): 212–224.

Thien, D. (2005) After or Beyond Feeling? A Consideration of Affect and Emotion in Geography. *Area*, 37(4): 450–454.

Tompkins, S. (1995) *Exploring Affect: The Selected Writings of Silvan S. Tompkins*. Edited by E. V. Demos. New York: Cambridge University Press.

Tulloch, J., Lupton, D., Blood, W., Tulloch, M., Jennett, C. and Enders, M. (1998a) *Fear of Crime Volume 1*, Canberra: Attorney-General's Department Commonwealth of Australia.

Tulloch, J., Lupton, D., Blood, W., Tulloch, M., Jennett, C. and Enders, M. (1998b) *Fear of Crime Volume 2*, Canberra: Attorney-General's Department Commonwealth of Australia.

Vanderveen, G. (2018) Gender, Violence and the Fear of Crime: Women as Fearing Subjects? In: M. Lee and G. Mythen (eds) *The Routledge International Handbook on Fear of Crime*. London: Routledge, pp 170–189.

Walklate, S. (2011) Reframing Criminal Victimisation: Finding a Place for Vulnerability and Resilience. *Theoretical Criminology*, 15(2): 179–194.

Walklate, S. (2018) Gender, Violence and the Fear of Crime: Women as Fearing Subjects? In: M. Lee and G. Mythen (eds) *The Routledge International Handbook on Fear of Crime*. London: Routledge, pp 222–235.

Warr, M. (1984) Fear of Victimization: Why Are Women and the Elderly more Afraid? *Social Science Quarterly*, 65(68): 1–702.

Warr, M. (1987) Fear of Victimization and Sensitivity to Risk. *Journal of Quantitative Criminology*, 3(1): 29–46. www.jstor.org/stable/23365514

Warr, M. (2000) Fear of Crime in the United States: Avenues for Research and Policy. *Criminal Justice*, 4: 451–489.

Wilson, J. Q. and Kelling, G. (1982) Broken Windows: the Police and Neighbor-Hood Safety, *Atlantic Monthly*, March: 29–38.

Wrenn, M. (2014) The Social Ontology of Fear and Neoliberalism. *Review of Social Economy*, 72(3): 337–353.

Young, J. (1987) The Tasks Facing a Realist Criminology. *Contemporary Crises*, 11: 337–356.

Young, A. (1996) *Imagining Crime: Textual Outlaws and Criminal Conversations*. London: SAGE.

Young, A. (2014) From Object to Encounter: Aesthetic Politics and Visual Criminology. *Theoretical Criminology*, 18(2): 159–175.

10

To Be a Homeless Woman in Russia: Coping Strategies and Meanings of 'Home' on the Street

Evgeniia Kuziner

Introduction

This chapter explores the affective strategies of homeless women and how they use buildings and public spaces in Russia to construct a 'home' on the street. Research suggests that there is an important gender dimension to the problem of homelessness (Edgar and Doherty, 2001; Watson, 2000) and that women's homelessness has remained largely invisible because of the particular stigma attached to the 'unaccommodated woman' – representative of a form of deviance, even if she may be simultaneously viewed as a victim and in need (Wardhaugh, 1999). Homeless women therefore disappear from the institutional spaces of homeless shelters and frequently rely on precarious arrangements to be housed (Pleace, 1998; Shinn, 2010). This chapter is based on research of homeless women conducted in St Petersburg. It explores women's affective strategies of coping on the street, place making and meanings given to 'home'.

Both men and women experience homelessness, but neither research, the media nor public policy pays enough attention to women's homelessness – the experiences of homeless women, the reasons for their homelessness and their strategies for surviving on the streets (Klodawsky, 2009). In the academic literature, some researchers argue that homelessness is predominantly male (Lenon, 2000; Alekseeva, 2003). In works that examine the historical context of homelessness, there is mention of the prevalence of the image of the homeless person as a 'bearded, dirty man' (Austerberry and Watson, 1983; Williams, 2001; Austerberry and Watson, 1981). According to Russian sociological research, the homeless are 'mostly men of working age who have served in the armed forces' (Alekseevna and Yuryevna, 2020: 14). However, the homeless are a heterogeneous social group inclusive of both men and women. Women's experiences, the reasons for homelessness, their practices of survival in a difficult situation are likely to be different from the experiences of men (Burt and Kohen, 1989; Bird et al, 2017). In particular, homeless women are more stigmatized, more vulnerable and more likely to be victims of violence.

Official data tends to give a misleading picture of the extent of homelessness in Russia. For example, the Russian national population census of 2021 shows that there are only 11,285 homeless 'households' in Russia, and only 32 homeless living in St Petersburg, the second-largest city. During the previous census, 10 years earlier, the number of homeless people was six times higher (64,077 in Russia and 2,902 in St Petersburg). One problem is that the census of 2021 does not include people who live in shelters or abandoned buildings. Further, in both censuses, officials use the term 'household' without specifying the number, gender or age of its members. One state agency has calculated that 3,352 people in St Petersburg are officially registered as homeless (to have the official status of homeless person is to not own any dwelling and come to the centre). However, the oldest Russian non-governmental organization (NGO) helping people experiencing homelessness, 'Nochlezhka' (based in St Petersburg and Moscow), states that there are approximately 64,000 homeless people in St Petersburg and 2.1 million overall in Russia (experts expect a dramatic increase in these numbers due to the COVID-19 pandemic and the ongoing war in Ukraine).

Sociological studies show that 80 per cent of homeless people in Russia are men (Sociological Portrait of a Homeless Man, 2008: 11). Homeless women are not only almost 'invisible' in statistics, they are also stigmatized. Historically, 'woman' and 'home' have been inextricably linked, and female homelessness has been seen as violating the social order (Oudshoorn, Berkum and Loon, 2018: 7). A woman deprived of her home was considered by society to represent an extreme degree of deviance, violating the order accepted by the state and the church (Casey et al, 2008). This raises the question of how, in the context of Russia, homeless women live, experience and cope with homelessness.

The role of affect in constructions of home

While men and women have similar perspectives on the definition of home in many ways, men are more likely than women to associate home with status and social achievement. Women, on the other hand, are more likely to see home through emotions as a place of safety and comfort (Sommerville, 1997), linked potentially to their traditional role as care givers. There is also evidence to suggest that women place a higher value on the home and derive greater satisfaction from it than men do (Mason, 1989).

The meaning of home can be examined through the integration of sociology of emotions with phenomenology of affected engagements (Sommerville, 1997), where concepts of privacy, identity and familiarity are key. This captures the physical layout of space as well as the nature of the psychological and social connections within it, where the interaction between self and others shapes personal and/or domestic identity. The latter requires a consistent and stable set of experiences, which includes familiarity with one's surroundings and the associated formation of boundaries. As a lived space of interaction and an expression or symbol of the self, homes can be seen as affective spaces in their capacities as centres of emotional significance, rootedness and belonging. This resonates with the concept of 'thick places' (Casey, 2001; Duff, 2010), which are inherently meaningful, constructed through affective engagement. Thick places are accordingly created through the intertwining of emotions, habits and significance, enhancing one's sense of belonging through feelings of affinity and the formation of strong emotional bonds.

The role of 'home' in determining the characteristics of women's homelessness

The importance of 'home' as a framework through which to examine the causes of homelessness has been much debated (Lawrence, 1995; Moore, 2007). As Somerville (1992) argues, the meaning of 'home' in the everyday lives of individuals is constructed by different actors such as the state, the media and the larger society. Here, the concept of 'home' is closely linked to the category of 'family' (Watson, 1984; Mallett, Rosenthal and Keys, 2005), which includes in its definition a sociocultural and historical understanding of a woman's role in the home as a 'hostess' (Passaro, 1996). Homelessness is therefore understood as the absence of a place where a person can spend time in a safe and private environment, alone or with family. From this perspective, home is conceptualized as a safe place, where one is free from fear and anxiety and can live independently without the influence of political and social processes (Kajka, 2004). As Moore (2014) argues, it is a place where a person experiences positive emotions and a sense of protection. However, feminist researchers have criticized this position, arguing that the home can also be a site of domestic violence (Malos and Hague, 1997). What may be a safe home for one person may be a place of oppression and violence for another (Wardhaugh, 1999). 'Home' therefore is a diverse concept with sometimes incompatible characteristics; for example, 'security' can border on 'danger' for different groups of people.

Although the concepts of 'home', 'family' and 'woman' are no longer so closely linked, this association has been seen as a reason why homelessness is most often associated with single men and why homeless women are often invisible and treated as a deviation (Mayock and Bretherton, 2016; Oudshoorn et al, 2018). This is particularly the case in Russia, where recent research suggests that homeless men seek help from NGOs 4.3 times more often than women (Bykova et al, 2023). This may reflect women's reluctance to enter male-dominated spaces, as well as a fear of victimization within services that are oriented primarily towards men. Women may also avoid seeking help or accommodation due to associated feelings of stigma or shame.

Everyday practices and coping strategies of women experiencing homelessness

Women use different practices and coping strategies to manage and survive homelessness. These can include sex work, using social networks and resources (social capital), using public spaces, finding a man or asking for help. Wesely (2009) has identified vulnerability and fear as determinants of women's everyday practices in managing homelessness. Homeless women are more likely than men to worry about their safety and to seek companionship in order to feel safe. As Gilfus (2006) claims, women may therefore use sex as one of their survival strategies on the streets. Fearing for their safety, they enter into relationships with men, exchanging sex for safety. Other researchers emphasize that these partnerships are often forced (for example, Casey et al, 2008). Relationships with men provide homeless women with safety from others, but it is common for women in such couples to be victims of violence (Gilfus, 2006) and/or alcohol abuse (Finfgeld-Connett, 2010). Other researchers studying the everyday practices of homeless women identify the use of public spaces as one of their coping strategies (Casey et al, 2008). Public spaces such as shopping centres, libraries and museums play an important role in the daily lives of homeless women, not only as places to wash, sleep and spend time in warmth, but also as places to maintain their status and differentiate themselves from other homeless people. According to researchers, by using public spaces, homeless women assert their right to be 'visible', to be like everyone else despite their situation. As Casey et al (2008) argue, homeless women's use of public spaces is not just a strategy for coping with homelessness but is also a strategy for combating homelessness as exclusion. This means that in order to use public spaces, homeless women have to play by certain rules: be invisible to security guards and be present during certain hours of the establishment.

Looking at the survival strategies of women living in shelters, Klitzing (2003) found problem solving, solitude and communication with others as key coping strategies. Problem solving consists of actions to 'get out of' the homeless situation, such as finding a job and/or acquiring key documents. In shelters where there is no space of one's own, being alone can be one of the everyday practices of homeless women. The researcher also highlighted religion and contact with religious organizations as a further coping mechanism. Denial of homelessness was another tactic. As Takahashi et al (2002) found, a woman can live in a shelter and still have a job, pay her bills and not consider herself homeless. Overall, research on women's homelessness needs to take into account both structural factors and the level of gender inequality in the country as a whole, as well as the individual woman's personal experience, her own perception of home and her role in it. At different stages of homelessness, women may use different survival strategies, including the use of 'female capital', social connections and resources as well as public spaces.

Methodology and empirical material of the study

The study, conducted between October 2018 and August 2021, adopted a qualitative methodology as the most appropriate for working in a sensitive field. Biographical interviews were conducted as the main research method, and data were collected through biographical semi-structured interviews. In total, 30 anonymous interviews were conducted which ranged from 50 minutes to two hours. The interviews were conducted in Russian, the audio transcribed and then translated by the author into English for full analysis. Supplementing the interviews was participant observation during volunteer work with several non-profit organizations, which also enabled me to make contact with potential participants, as well as visits to abandoned houses. For example, observations were made while giving food to the homeless in several locations in St Petersburg and while working with volunteers providing first aid. Several observations were also made in abandoned houses where groups of the homeless were living. Overall, the project's sample comprised women living in shelters in the city, women who have experienced homelessness and are currently renting accommodation and women living on the streets of the city. I did not interview women with a 'hidden' form of homelessness, namely, those who had a problem with housing but who were living with relatives, friends or acquaintances.

The home that doesn't exist: everyday practices of homeless women

Women with experience of street life have often slept in basements, attics and abandoned houses. In the following, the chapter will talk about those women who live or have lived 'on the street' and not in shelters run by charities. In this case, 'home' is a physical place, a dwelling that needs to be cared for and furnished. The 'ideal' place to live is, in principle, similar to what is seen in 'normal' domestic accommodation, namely, hot and cold running water, electricity or the possibility of it, the possibility of going unnoticed by other occupants and the absence of other homeless people. In most cases, women sought accommodation with a partner. Due to the small sample size, it is difficult to judge whether this choice is due to the partner's experience or to the fact that in these cases men play the role of caretakers and can act from a position of strength (for example, kicking out strangers) if necessary.

'No, of course I didn't, S. (male partner) did. He knew all the local basements, because he used to spend the night there with a (name of a substance), he would drink there, he would run around in the basements, you can't go down the street like that. Well, they're all the same, here, I don't know any attics or basements.' (Interview No 3, Natalya, 19 years of homelessness, living in a shelter at the time of the interview)

The ideal number of inhabitants of a basement or attic of a residential building should be no more than three or four. The more people, the more likely you are to be spotted by the inhabitants of the building. If you are spotted, you will probably have to leave and look for a new place to live.

'We are much better together. First of all, it depends on the conditions. The company is just a way to relax. It's during the day. We were talking anyway. During the day you can sit there as a group. There, somewhere on a clearing, to go somewhere. But at night, to go into the front room [of the basement], it's better with two of us. Much better. First he warns me, then I warn him. Where it's quieter, where what's what, and if you crowd like this ...' (Interview No 15, Olga, 67 years old, 19 years homeless, living in a shelter at the time of the interview)

In abandoned houses, the number of tenants can be higher, depending on the size and number of rooms. For example, in a large, abandoned house where the research took place, there were more than ten homeless people at the time of the study: several couples, several singles and a group of several men who lived together. All lived in isolation, in separate rooms, with a toilet and a common dining room, and to enter a room one had to knock. Despite the segregation, there was a practice of celebrating holidays and eating together. There was a very well-developed communication network: if someone moved in at the other end of the house, everyone would know, usually within a day.

There were, however, strict rules about appropriate behaviour, particularly the need for quiet (if it is a basement or attic) and cleanliness. Breaking these rules is critical for the whole group, as making noise can attract the attention of tenants, who may call the police and evict the homeless.

'They spoiled everything. They were screaming. There was a girl, she was 20 years old. She used to fight with her guy, swore. She didn't behave like – come, sit at home, don't go anywhere. To enter our basement, you should go through the main entrance, people see it. Here they begin to fight, swear, squeals, squeaks. And that's it. Everything is heard in the entrance.' (Interview No 2, Nina, 42, homeless for 17 years, living in an abandoned house at the time of the interview)

In order to be invisible, homeless people living in abandoned houses settle as far and deep inside the house as possible – the harder it is to get to the 'inhabited' floor, the better. To get to the homeless living in an abandoned house, my guide and I had to climb through a small window hidden behind a protective green mesh (the house is in a state of disrepair), climb half-destroyed stairs with no railings and go around an old fridge in the middle of the hallway, leaving only a small passage by the wall, not big enough for a grown man to pass through. In other words, to get to these homeless people required a lot of effort and knowing exactly which floors they lived on. In another abandoned house, there was a whole warning system – a rope with cans and bottles stretched across the floor as an alarm when someone unknown passed by.

When describing their homes, many of the informants focused on the usual everyday things that make their home seem 'normal', different from the stereotypical accommodation of a homeless person. In homeless families, the woman is usually in charge of daily life, while the man's role is to generate electricity, set up a washing system if possible and provide security. The woman usually cooks and cleans.

'No, no, they came to visit, but not like that … I know, there are a lot of such acquaintances who live together – ten people or more. No, we lived two of us. We sat there with company and shared food, drank, smoked, all like that. That is, no one spent the night with us. I had a stove, we had a refrigerator. … And there is water, light, both hot and cold water, of course. There was a TV, a tile, too, we had a bed there, I cooked, everything was the same as at home, just yes, here are the pipes, we had a little room there … that is, everything that was necessary.' (Interview No 3, Natalya, 51 years old, experience of homelessness 17 years, living in a shelter at the time of the interview)

It is interesting how this passage deals with the symbolic separation of the self from others, the 'homeless'. The ordering and normalization of the home space in the form of cleanliness and hygiene is the basis for this separation, helping to create a domestic identity rather than a homeless one. We can distinguish the following strategies of normalization: zoning the space, furnishing the space with furniture and appliances, introducing pets, observing the norms accepted for domestic conditions (for example, wiping one's feet when entering). The space is zoned as in a normal house: there are separate bedrooms, a common place, namely the living room, a separate place for the toilet. In interviews and discussions, the women emphasized that the creation of such living conditions makes them 'ordinary people'.

'No, it was neat in my room, I swept, that is, I threw papers in the same way as ordinary people throw them in a garbage bag. But not like, you know, since you are in the basement, then you can throw it like that, and shit, and piss where you live. No. We have a separate toilet, we used it. In this regard, everything was neat with us. I washed clothes – the same as at home, everything was there.' (Interview No 3, Natalya, 51 years old, experience of homelessness 17 years, living in a shelter at the time of the interview)

It is the zoning of the room that gives a woman more privacy and her own space. For example, Nina lived with her partner in a separate room, she had a separate space for cooking and a dressing table with cosmetics. Stating, "I don't use makeup at all", it was still important for Nina for her to conform to what she saw as a 'normal' woman. Furnishing the room with furniture and appliances (often non-working) creates a sense of a real 'home' and comfort. One woman had once brought a TV with a broken screen into the basement where they lived, put it in the middle of the room, like in normal living rooms, and used it as a lamp. Pets are also a sign of domesticity and livability, helping to normalize the household. As Nina continued: "We lived in the basement on (name of the district), we had a little kitten, he only slept with me, he loved only me" (Interview No 2, Nina, 42, homeless for 17 years, lived in an abandoned house at the time of the interview).

Hygiene, wearing clean clothes and the use of cosmetics are important practices in the lives of homeless women, helping to create what they see as a 'normal' home and creating distance from a homeless identity. When talking about the places where they had lived, one of the advantages that the women immediately mentioned was the availability of hot and cold water, that is, the possibility to bathe regularly.

'You see, those homeless people who come in the morning, I'm shocked by them. That is, of course, I compare myself, I didn't go so shitty and dirty. Dumps were built here, we dressed from the dump, that is, people in the store asked me: "Natasha, where do you get clothes?" I answered: "From the garbage dump."' (Interview No 3, Natalya, 51 years old, experience of homelessness 17 years, living in a shelter at the time of the interview)

If there is no running water in the area where the homeless live, water is taken from toilets in shopping centres, petrol stations and other public places. You can also clean up there. Looking tidy is a resource that can help you get help. If you don't look or smell like a homeless person, people are more likely to help you and generally treat you better. Clean clothes make it easier to get a good job and access public places.

Due to difficult living conditions, poor nutrition and stress, most homeless women do not menstruate. If the menstrual cycle continues, they use means such as rags and bandages. Hygiene products are too expensive and are used only if they come from 'outside', as a donation. Talking about children is also a very sensitive subject, as women living on the streets or in shelters do not have the opportunity to live with their children. Several informants had become pregnant while living on the streets. Marina and Nina's newborn children were taken away from the maternity hospital. Marina has relatives in another city, her homelessness is sporadic, she wants her baby back and talks about the situation all the time. Nina, who has been living in an abandoned house for a long time and has no ties to the outside world, has treated the situation as something she cannot change: "I was told – we won't give him back, you don't have a registration, you have nothing, where will you take him?" (Interview No 2, Nina, 42 years old, 17 years of homelessness, living in an abandoned house at the time of the interview).

'Home' markers have limitations: you can't paint or wallpaper the walls, especially if the accommodation is in an attic or basement. A homeless person's 'home' has a life span – usually no more than a year or two, depending on the location (abandoned houses last longer than attics or basements) and the relationship with the neighbours who live there.

Conclusion

This chapter has outlined the attempts made by homeless women to recreate a home in the conditions available to them. They make use of derelict premises that were not originally intended for habitation and premises that were once occupied. These may include makeshift huts, though these are short lived and can be vandalized, which makes them unsafe. With their partners, women strive to make a 'real' home by installing electricity, connecting the house to the water supply and finding basic items such as beds, crockery and cookers. At the same time as a 'home' is being built, personal boundaries are being created between the 'home' and the families living there. Homeless people visit each other, bringing with them the obligatory attributes of friendly gatherings: alcohol, sweets and other gifts.

Women strive to create a space that is as close as possible to what they see as a 'normal' home. This may include a television that does not work but is seen as essential to a normal domestic space, or a dressing table that is not used but which holds meaning as a 'feminine' artefact in other women's daily lives. Homeless people may have animals that are treated as pets. On an emotional level, women cope with homelessness with its insecurity and precarity by creating spaces of privacy (separated rooms for couples), by distancing themselves from an identity of homelessness and by making their temporary accommodation as close as possible to what they see as a secure and permanent home. Here, we can see how the meanings attached to home reflect a 'feeling state' that is generative of action through the various coping strategies outlined earlier. These conform to 'thick places' (Casey, 2001) that are made out of affect and practice and which enhance a sense of belonging. These spaces, despite their temporary nature and the constant threat of eviction and disturbance, offer emotional comfort and help to create a valued 'domestic' identity.

References

Alekseevna, T. E. and Yuryevna, K. N. (2020) Social and Legal Aspects of the Problem of Homelessness in Russia. Based on the Materials of an Interregional Study, *Monitoring Public Opinion: Economic and Social Change*, 2(168): 409–429 (in Russian).

Alexeeva, L. S. (2003) Homeless in Russia. *Social Research*, 9: 56–62.

Austerberry, H. and Watson, S. (1981) A Woman's Place: A Feminist Approach to Housing in Britain. *Feminist Review*, 8(1): 49–62. https://doi.org/10.1057/fr.1981.11

Austerberry, H. and Watson, S. (1983) *Women on the Margins: A Study of Single Women's Housing Problems*. London: Housing Research Group.

Bird, M., Rhoades, H., Lahey, J., Cederbaum, J. and Wenzel, S. (2017) Life Goals and Gender Differences Among Chronically Homeless Individuals Entering Permanent Supportive Housing. *Journal of Social Distress of Homeless*, 26(1): 9–15. https://doi.org/10.1080/10530789.2016.1274570.

Burt, M. R. and Cohen, B. E. (1989) Differences Among Homeless Single Women, Women with Children, and Single Men. *Social problems*, 36(5): 508–524.

Bykova, A., Filippov, N. and Yamshchikov, I. P. (2023) Rehabilitating Homeless: Dataset and Key Insights. *Proceedings of the AAAI Conference on Artificial Intelligence*, 37(12): 14136–14143. https://doi.org/10.1609/aaai.v37i12.26654

Casey, E. (2001) Between Geography and Philosophy: What Does It Mean to Be in the Place-world?' *Annals of the Association of American Geographers*, 91: 683–693.

Casey, R., Goudie, R. and Reeve, K. (2008) Homeless Women in Public Spaces: Strategies of Resistance. *Housing Studies*, 23(6): 899–916. https://doi.org/10.1080/02673030802416627.

Duff, C. (2010) On the Role of Affect and Practice in the Production of Place. *Environment and Planning D: Society and Space*, 28(5): 881–895. https://doi.org/10.1068/d16209

Edgar, B. and Doherty, J. (2001) *Women and Homelessness in Europe. Pathways, Services and Experiences.* Bristol: The Policy Press.

Finfgeld-Connett, D. (2010) Becoming Homeless, Being Homeless, and Resolving Homelessness among Women. *Issues in Mental Health Nursing*, 31(7): 461–469. http://dx.doi.org/10.3109/01612840903586404.

Gilfus, M. E. (2006) From Victims to Survivors to Offenders: Women's Routes of Entry and Immersion into Street Crime. In: L. Alarid and P. Cromwell (eds) *Her Own Words: Women Offenders' Views on Crime and Victimization.* Los Angeles, CA: Roxbury, pp 5–14.

Kaika, M. (2004) Interrogating the Geographies of the Familiar: Domesticating Nature and Constructing the Autonomy of the Modern Home. *International Journal of Urban and Regional Research*, 28(2): 265–286. https://doi.org/10.1111/j.0309-1317.2004.00519.x.

Klitzing, S. W. (2003) Coping with Chronic Stress: Leisure and Women Who Are Homeless. *Leisure Sciences*, 25(2–3): 163–181. https://doi.org/10.1080/01490400306564.

Klodawsky, F. (2009) Home Spaces and Rights to the City: Thinking Social Justice for Chronically Homeless Women. *Urban Geography*, 30(6): 591–610. https://doi.org/10.2747/0272-3638.30.6.591.

Lawrence, M. (1995) Rural Homelessness: A Geography without a Geography. *Journal of Rural Studies*, 11: 297–307. https://doi.org/10.1016/0743-0167(95)00025-I.

Lenon, S. (2000) Living on the Edge: Women, Poverty and Homelessness in Canada. *Canadian Woman Studies*, 20(3): 123–126.

Mallett, S., Rosenthal, D. and Keys, D. (2005) Young People, Drug Use and Family Conflict: Pathways into Homelessness. *Journal of Adolescence*, 28(2): 185–199. https://doi.org/10.1016/j.adolescence.2005.02.002.

Malos, E. and Hague, G. (1997) Women, Housing, Homelessness, and Domestic Violence. *Women's Studies International Forum*, 20(3): 397–409. https://doi.org/10.1016/S0277-5395(97)00023-X.

Mason, J. (1989) Reconstructing the Public and the Private: the Home and Marriage in Later Life. In: G. Allan and G. Crow (eds) *Home and Family: Creating the Domestic Sphere.* London: Palgrave Macmillan UK, pp 102–121.

Mayock, P. and Bretherton, J. (eds) (2016) *Women's Homelessness in Europe.* London: Palgrave Macmillan. https://doi.org/10.1057/978-1-137-54516-9.

Moore, J. (2007) Polarity or Integration? Towards a Fuller Understanding of Home and Homelessness. *Journal of Architectural and Planning Research*, 24(2): 143–159.

Moore, R. (2014) Coping with Homelessness: An Expectant Mother's Homeless Pathway. *Housing, Care and Support*, 17(3): 142–150. https://doi.org/10.1108/HCS-02-2014-0002.

Oudshoorn, A., Berkum, A. V. and Loon, C. V. (2018) A History of Women's Homelessness: The Making of a Crisis. *Journal of Social Inclusion*, 9(1): 5–20. http://doi.org/10.36251/josi.128.

Passaro, J. (1996) *The Unequal Homeless: Men on the Streets, Women in Their Place*. New York: Routledge.

Pleace, N. (1998) Single Homelessness as Social Exclusion: The Unique and the Extreme. *Social Policy & Administration*, 32: 46–59. https://doi.org/10.1111/1467-9515.00085.

Shinn, M (2010), Homelessness, Poverty and Social Exclusion in the United States and Europe. *The European Journal of Homelessness*, 4: 19–44.

Sociological Portrait of a Homeless Man (According to the Data of the Interregional Study 'Legal and Social Aspects of the Problem of Homelessness', 2006) (2008) Homelessness in Modern Russia: Problems and Ways to Solve Them. *Bulletin of the Interregional Network 'For Overcoming Social Exclusion*, 1. Saint Petersburg (in Russian).

Somerville, P. (1992) Homelessness and the Meaning of Home: Rooflessness or Rootlessness? *International Journal of Urban and Regional Research*, 16(4): 529–539.

Somerville, P. (1997) The Social Construction of Home. *Journal of Architectural and Planning Research*, 14: 226–245.

Takahashi, L., McElroy, J. and Rowe, S. (2002) The Sociospatial Stigmatization of Homeless Women with Children. *Urban Geography*, 23(4): 301–322. https://doi.org/10.2747/0272-3638.23.4.301.

Wardhaugh, J. (1999) The Unaccommodated Woman: Home, Homelessness and Identity. *The Sociological Review*, 47(1): 91–109. https://doi.org/10.1111/1467-954X.00164.

Watson, S. (1984) Definitions of Homelessness: A Feminist Perspective. *Critical Social Policy*, 4(11): 60–73. https://doi.org/10.1177/026101838400401106.

Watson, S. (2000) Homelessness Revisited: New Reflections on Old Paradigms. *Urban Policy and Research*, 18(2): 159–170. https://doi.org/10.1080/08111140008727830.

Wesely, J. K. (2009) 'Mom said we had a money maker': Sexualization and Survival Contexts among Homeless Women. *Symbolic Interaction*, 32(2): 91–105. https://doi.org/10.1525/si.2009.32.2.91.

Williams, M. (2001) Complexity, Probability and Causation: Implications for Homelessness Research. *Journal of Social Issues*, 1(2). www.whb.co.uk/socialissues/mw.htm (accessed 17 August 2020).

PART III

Place, Gender Identity and Belonging

11

Affective Atmospheres of Finance: Gendered Impacts of Financialization within Sydney's Barangaroo Development

Alex Simpson and Paul McGuinness

Introduction

To stand beneath the towering steel-and-glass edifices of Sydney's latest commercial development, Barangaroo, is to witness – and feel – the smoothed-out, friction-free flows promised by contemporary financial capitalism. Formally known as East Darling Harbour and situated on the unceded lands of the Gadigal people of the Eora nation, Barangaroo is a key pillar of the Global Sydney project, built with the intent of attracting the interests of capital and shaping the future of Sydney's landscape accordingly (Baker and Ruming, 2015; Harris, 2018). Spanning 22 hectares of previously industrial land to the west of the Harbour Bridge, Barangaroo is the largest urban redevelopment project in Sydney since the 2000 Olympics and was announced by the then New South Wales premier, Bob Carr, as a 'historic opportunity to return a substantial part of Sydney's foreshore back to the people' (NSW Government, 2005). However, as the development reaches its final phase, Barangaroo is increasingly proving to be a 'placed' landscape, built for – and by – the 'progressive' and 'virtuous' allure of finance. Populated by premium financial firms, a scandal-plagued casino and the homogenized blur of white faces, dark suits and crisp shirts, Barangaroo stands as an isolated cluster removed from the surrounding areas. Like so many urban regeneration projects, Barangaroo is designed to create a space that transforms its host city into a key strategic hub of global financial exchange (Harris, 2018; Burns, 2019). Walk along the slicken wooden promenade, the stains of labour and history have been removed, instead the curated presence of manicured waterfront eateries, designer shops and financial institutions, all swarmed by beaming faces, slip into and become one to imbue this space with the affective atmosphere of place and, with it, belonging.

If place making is about creating and sustaining an imaginary, giving form to a non-representable idea (Rancière, 2007) or, more precisely, giving 'figurative shape to presence' (Abraham and Torok, 1972: 128), then affective atmospheres describe the impact such force of presence has on the bodies who occupy (or are excluded from) space (Böhme, 1993; Brand, 2023). The affective atmospheres of finance, in other words, bring into sharp focus the dynamic, productive and felt interplay between spatiality, affect and assembling of bodies, emotion, discourse, materiality and technology (Brand, 2023). After all, this milieu of financial productivity gives rise to a 'character' of spatial form; one which, when enveloped within a broad sensory spectrum, can be described as an 'atmosphere' that produces and responds to the multiple expressions of spatiality. Bissell (2018: 272) suggests that 'affective fields' reflect how people, buildings and technologies, including forms of non-human life, come together in a particular geographical setting. These sensory, emotive and embodied affections in response to place create, in turn, an 'energy of feeling' that may arise but dissipate, depending on the context of environmental emersion: an affective atmosphere (Anderson, 2009; Young, 2019). As such, affective atmospheres are central to everyday conduct. Yet, important here is how different atmospheres, specifically financial atmospheres, facilitate and restrict particular practices and, in doing so, precipitate particular structures of gender, emotion and feeling.

As Anderson (2009) reminds us, spatial practices as diverse as interior design, landscape gardening and architecture all aim to produce and circulate a distinct atmosphere. By creating and arranging material fragments of light, sounds, symbols and texts, atmospheres are 'enhanced', 'transformed', 'intensified', 'shaped' and otherwise impacted upon to create affective qualities of homely, serene, welcoming and so on (Böhme, 2006; Anderson, 2009). It is not just that each of these characteristics carries discursive constructions of gender, but the mood that is elicited, as well as the impact and emotive response, shapes understandings (and feelings) of belonging. In other words, if an atmosphere, as Brennan (2004: 1) suggests, 'literally gets into the individual', then the production and encounters of atmospheres need to be inclusive of gender as a prism that shapes our encounters with atmospheres of being in/out of place. With this in mind, detailed attention is paid to the textures, spatial arrangement and material topography that characterize the social and structural organization of Barangaroo. As Young (2019) argues, spatiality, affect and the material aesthetics of space conjoin in atmospheres, shaping our sense of belonging, identity and affinity within any given locale. For Stewart, this means, the characteristic of the spatial form and 'affective qualities emanate from the assembling of the human bodies, discursive bodies, non-human bodies and all the other bodies that make up everyday sites' (Stewart, 2007: 80). Much attention has been given to the psychic and emotive interaction between bodies, on the one hand, and the placed environment, on the other. Yet, atmospheres can surreptitiously, insidiously, resist the incorporation of gender or other markers of identity and difference. This is despite the noted grip of anxiety or the ease of belonging that atmospheres can evoke; a sense that, in very real terms, shapes possibility, movement and perception (Anderson, 2009; Frosh, 2012; Young, 2019).

Drawing on these debates, and focusing on the relationship between atmospheres, gender and belonging in Barangaroo's financial arena, this chapter examines the affective atmospheres of finance and its relationship with place making. Finance is, after all, not only placed but a material endeavour with deep ,gendered relations (McDowell, 1997; Sassen, 2005). In examining the affective atmospheres, we view Barangaroo as a figurative shape that contains the affective relations of presence, identity and emotion, which are woven into the material and affective atmosphere of place and that give meaning to core concepts of gendered belonging. In architecture and urban design, we can see the incorporation of atmospheres manifest in the physical form: a materiality wrapped around an aspiration that is as much about reshaping a new reality as escaping an old one (Sudjic, 2005; Spencer, 2016, 2018). As we argue in this chapter, the construction and maintenance of Barangaroo – like so many other financial centres – is about introducing the structuring logic of finance into the material, cultural and social form, creating the perfect conditions of the idealized global financial marketplace, where the movement of knowledge and financial transactions are unencumbered by the latency of slow decision making, decaying bodies or imperfect systems. In doing so, we focus on and explore the affective atmospheres of the boundary, temporality and the slippery material properties of glass and light, as well as the gendered effects of these spaces. The outcome is a sensory dislocation, neither being solid nor ephemeral, floating within a temporality of its own making that propagates gendered bodily struggles with decay, frailty and permeability. As we argue, this tough, exclusionary exoskeleton that surrounds financial place-making projects, such as Barangaroo, engenders a sense of alienation and displacement by embodying or projecting finance capital's culture of masculinist rationality.

Affective atmospheres of place

The process of place making and urban design has long been associated with 'atmosphere'. Not only located within spaces that contain a rich narrative history, such as cathedrals or national monuments, atmospheres can be seen to permeate through the affective formation of space (Bissell, 2018;

Brand, 2023). So, what is an atmosphere? We may think of atmospheres as 'spatialized feelings' (Griffero, 2014: 37) or 'emotionally felt spaces' (Böhme, 1993: 114), which points towards the ways in which place and our interaction with our material surrounds elicits an ever-shifting emotive and embodied response (Brand, 2023). Yet, as Bissell (2018) argues, affective atmospheres are perceived and sensed through the body. As Urry (2007: 73) clarifies, 'atmosphere is in the relationship of peoples and objects'. It is something sensed often through movement and experienced in a tactile kind of way. Here, atmospheres work between space and are woven into body discourse and the body to create an often pre-reflexive experience of space before we even set foot in it. Above all, atmospheres are affective: emerging through and out of the space in between – in between our own subjective experience and the material, social and technological tapestry that surrounds us (Thrift, 2004; Gregg and Seigworth, 2010). As Anderson states, affective atmospheres are 'a class of experience that occur *before* and *alongside* the formation of subjectivity, *across* human and non-human materialities, and *in-between* subject/object distinctions' (Anderson, 2009: 78, emphases original). It would be wrong, therefore, to associate atmosphere with any singular subjective reaction to place (Young, 2019); rather, atmospheres are always located in between experiences and environments (Bille et al, 2015) and should be conceptualized as a relation of felt connections that both shape and are shaped by the surrounding environment (Anderson, 2009).

Cities and the placed environment are often seen to be inhuman or transhuman, which, rather than fostering communities and community, function as little more than containers in which the politics of socialization takes place (Thrift, 2004). To frame places as affective is to focus on the rolling maelstrom of emotions and, more aptly, atmospheres that cities engender. Atmospheres – like places – may interrupt, perturb and haunt individuals, places or things. A focus on *affective atmospheres*, therefore, begins to foreground the shared, placed environment from which our individual and collective subjectivities, feelings and emotions arise – be it a sense of fear or belonging, joy or despair (Anderson, 2009; Brand, 2023). Here, the formation of place weaves through the material and social surrounds to locate, within us, a sense of presence – formed around an axis of inclusion and exclusion – which can be framed as 'atmospheric'. Young (2019) talks about the atmosphere that exists between the prison guard and the prisoner, along with the spatial framing that shapes this relationship, while Frosh (2012) describes how atmospheres exist as a force of feeling: a feeling that skips across space to transfer registers of fear, joy or uncertainty from person to person, site to site, as a collective shiver or expression of mood. This *transference* of thought, feeling and emotion is rooted in space, giving rise to an invisible sensation in which we each find ourselves troubled by the objects around us as well as, more crucially, each other (Frosh, 2012: 243). Atmospheres are, therefore, a 'spatially extended quality of feeling', which is both 'absorbed into a bodily state of feeling' and 'fills space with a certain tone' (Böhme, 1993: 117–118). As such, atmospheres are viscerally felt, yet distributed through the spatial frame to register a diffuse bodily and emotional sense (McCormack, 2008), which 'seem to fill the space with a certain tone of feeling like a haze' (Böhme, 1993: 114).

While atmospheres may be invisible and exist as only a felt force, they form a ubiquitous and constant backdrop to our everyday life as we move in, out and through space. These atmospheres, as Bissell (2018) argues, affect the ways in which we inhabit space and, for Edvardsson et al (2003), define what we must *cope with* and suggest direction of *behaviour*. While much focus has been placed on the spatial characteristics of atmospheres (for example, Böhme, 2006; Bille et al, 2015; Brand, 2023), it is important to examine the gendered process of what Anderson (2009: 80) calls 'envelopment' and the divergent impression atmosphere makes upon our gendered selves. As both a sensory force and a material presence, atmospheres become useful to think with because they hold a series of opposites: presence and absence, materiality and ideality, definite and indefinite (Anderson, 2009: 80). These tensions that these binaries create shape how we internalize and respond to the world around us, and our embodied (and emotional) response, giving meaning to routines, culture and belonging. But it is important to stretch these binaries to incorporate the multiple dynamics of inclusion and exclusion, which include, among other markers of identity, gender, race and/or sexuality. Truly progressive urban futures would decentre 'white, able-bodied, cis man as the default subject and imagine everyone else as a variation to the norm' (Kern, 2021: 54). Barangaroo being born of finance capital makes such a vision impossible, a lost future. While cities such as Seoul and Vienna have adopted 'gender mainstreaming' practices into their urban planning strategies, Kern (2021: 49) warns that the imagined beneficiary is a married, able-bodied mother with a pink- or white-collar job, excluding the likely majority of women who do not fit this description in most contemporary cities. As atmospheres are never static, but in the constant process of emerging and transforming (Bille et al, 2015; Young, 2021), greater framing is needed in relation to the gendering of atmospheres –in terms both of the spatial dynamics and the embodied response, as well as all that is in between. It is important to state here that atmospheres are not simply what we make them. Rather, atmospheres exist within the folds and gaps of ambiguity, working through the world around us in a way that may be felt, but without being instantly reducible to any material (or bodily) source (Anderson, 2009; Young, 2019). Yet, the force of feeling, that spatial coding and the impacted experience contain markers of gendered experience and sensemaking.

By devoting so much square footage to the valorization of finance capital, Barangaroo stands as a monument to gender inequality. Finance remains a resolutely macho profession, especially in its upper echelons: The Official Monetary and Financial Institutions Forum (OMFIF) found that across 336 of the largest central banks, commercial banks, pension funds and sovereign funds only 14 are led by women and it will take 140 years to achieve gender balance at current rates of change (OMFIF, 2023). Barangaroo is an abstract space striving to be thought of as 'place'. Its form seduces visitors with a spectacle that evokes progress, inclusivity, openness and a politically sustainable future. Yet, this future is illusory. Just like the financial abstraction that it was built for, Barangaroo's attempted materiality can be grasped only abstractly. Its coherence is revealed to be falsified when more closely examined. Financial centres like Barangaroo aim to make identity, including gender, abstract. Sassen (2019) identifies how these spaces denationalize the corporate elite – reducing sociality to a means of maximizing and managing the speeds wrought by technological connectivity and softening the human stress associated with its speed. Inclusion in abstract space is an illusion, for, as Lefebvre (1991: 287) states, it 'is not homogeneous; it simply has homogeneity as its goal'. By seducing visitors with its projected benevolence, Barangaroo masks its foundations in financial abstraction, embodying 'a successful integration of spectacle and violence' (Lefebvre, 1991: 307). The human lack of abstract spaces is offset by the 'phallic verticality' of the skyscraper, 'ensuring that "something" occupies this space, namely, a signifier which, rather than signifying a void, signifies a plenitude of destructive force – an illusion, therefore, of plenitude, and a space taken up by an "object" bearing a heavy cargo of myth' (Lefebvre, 1991). Weisman (1981: 6) describes the skyscraper as the 'pinnacle of patriarchal symbology', the full balloon of the inflated masculine ego that dominates city life, its fetishized 'procreant power' perpetuated in the architectural vernacular of 'base', 'shaft' and 'tip'. Hayden (1977) reminds us that the archetypically American skyscraper is but the latest iteration of phallic militaristic monuments throughout history – poles, obelisks, spires, columns, watchtowers preceding – where the imagery of war and patriotic death was overlaid with the corporate imagery of fecundity and generative power.

Traversing atmospheric and temporal boundaries

While, for much of the time, atmospheres are almost imperceptible, filtering through the affective layers of material and social organization that gives meaning to space, there are moments when atmospheres come to the surface and are more apparent (Anderson, 2009; Young, 2019; Brand, 2023). Through this framing, boundaries, or markers of transition, are useful devices to think through. As Lamont and Molnar (2002) argue, a study of boundaries holds the power to shed new light on the strong axis of creation and the force of relations that come to create gendered understandings of belonging as well as a demarcated sense of being. Boundary work, therefore, begins to reveal atmospheric markers of difference, through which sensory processes of belonging or exclusion can emerge as groups differentiate themselves from others by drawing on criteria of community and shared belonging (Lamont and Molnar, 2002; Bourdieu, 2010). In turn, the ability to draw and enforce a definable boundary leads to a distinct formation of social identity, knowledge and spatial community that holds the power to include and exclude in equal measure. The boundary, therefore, is a powerful imaginary that elicits a palpable, yet not always visible, atmospheric presence that holds the power to define and demarcate, exclude and incorporate.

Focusing on Barangaroo, its placement is a substantial part of the tapestry of Sydney's evolving central business district (CBD), yet it stands both within its grasp but also apart. Separated by the Western Distributor, the main traffic artery that flows from the Harbour Bridge and through the heart of the city, Barangaroo sits to the previously uncommercialized west of the harbour. Bridges crisscross the maze of elevated and street-level thoroughfares, linking these two commercial developments in a disarticulated, post hoc manner. The separation here only adds to the sense of 'entry' that comes with Barangaroo: a sensory feeling that atmosphere, design and articulation of space are perceptibly different from that of the traffic, noise, fumes and strangled post-war road planning that bifurcates the City of Sydney. It is an ancillary that lines and faces the glistening waters of Darling Harbour and which is connected through the corporate umbilical cord of Wynyard Station (Figures 11.1 and 11.2). As one of Sydney's main business thoroughfares, Wynyard traditionally faced east, funnelling out corporate workers towards the traditional financial centre of George Street and Martin Place. Now, with a revamp, the new corridors of traffic filter west towards Barangaroo. Traversing the void that separates the CBD from this new development, you are reminded of Young's (2019: 767) words: 'Atmospheres exist most noticeably as a phenomenon that is encountered: most obvious in the initial moments of an encounter, receding or diffusing as it becomes familiar or expected.' It is in this moment, stepping over the often invisible but atmospherically perceptible boundary, that an atmosphere of place is first felt and experienced, before this feeling is lost and blurs into the background of sensory expectation.

These affective ties, as Ahmed explains, 'involve subjects and objects, but without residing positively within them' (Ahmed, 2004: 119). It is, in other words, a new, shimmering world of lofty, glass structures, smoothed-out concrete services and fledgling trees which promise to expand their canopy, yet remain trapped within its very much urban surrounds (Figure 11.3). This mixture of the 'soft' of nature encased within the 'hard' materiality of urban lends a sense of concrete expansion, urban colonization. The sharp reflection of the glass looms up ahead, refracting the unfiltered

Figure 11.1: Bridge across Sussex Street from Wynyard Station

Source: Alex Simpson

Sydney sun, and stands as a mirror that rebounds the world around. More is discussed about the function of glass in the next section, but this is, very much a sociality with determinate borders, where the division of space, its patrolled entry and coalescence of wealth are written, visible through the dominant material tapestry. These affective relations generate collectives that envelop and reflect changing material surfaces, technological apparatus and the social circuits of infrastructure that enable a sense of place and belonging to occur (Bissell, 2018). It is a material space that, thinking through the relations between bodies, holds the affective capacity to bring people together, yet continues to demarcate, in Sack's (1993: 326) terms,

Figure 11.2: View across Sussex Street to Barangaroo

Source: Alex Simpson

between *guest*, *stranger* and *citizen*. To pass through the slick, fluid, yet reinforced security that encloses financial architecture is to instil a sense of belonging, or *citizenship*, among its members (Simpson, 2022). Its spatial and atmosphere formation, as Threadgold (2020) argues, helps to reaffirm a broader *field of power* to create a material environment that generates a social selective sorting of people and practices within the spatial framework of, in this case, Barangaroo.

Surrounding the new, glistening glass edifices, now ubiquitous to financial landscaping, the immediate feeling is one of familiarity – Barangaroo could

Figure 11.3: Urban garden in front of the International Towers

Source: Alex Simpson

be almost, without exception, any financial centre. The smoothed-out, friction-free surfaces (Spencer, 2016, 2018) evoke the dual solidity and ephemerality of finance: at once solid and forever shifting. This is, quite simply, a space built for capital and by capital. In doing so it privileges embodied power relations, historical orthodoxies, customs and rituals of financial capitalism, reifying a gendered spatial environment that marks 'belonging' through a shared elective membership of financial capitalism. As Noble (2013: 355) argues, 'fields are not simply objectified social spaces, but virtual spaces we carry with us in our embodied, socially shared capacities activated in institutions, occasions and settings'. In other words, the relationship between Bourdieusian fields and places is the ways in which the shared institutional settings are woven through the material tapestry of space, bringing to life a common affinity that marks 'belonging' across financial architecture – from Barangaroo to Singapore, Wall Street and the City of London. Each is connected materially, technologically, but also atmospherically. Here, just as finance is deeply gendered and, as Sylvia Walby (2015) reminds us, woven into the masculine neoliberal project, then so too is its spatial surrounds.

Bringing an affective layer to thinking about Bourdieu's field, Threadgold (2020) imagines connected but separates space as having a conjoined affective atmosphere and 'structures of feeling'. Structures, histories, norms, traditions and belonging are, after all, both aspects that give relational meaning to a field but also a collection of affects with their own gendered hierarchy of relations. Imagining Barangaroo in this way emphasizes how space is both rooted in material presence as well as capable of transcending ontological frames to articulate an ever-present affective atmosphere that brings to life a shared sense of being and markers of (gendered) identity.

Sensing the 'feel' of a place, being attuned to the atmospheric presence, is about understanding the unsaid assumptions and rules active within the environment and being able 'to anticipate what comes next' (Threadgold, 2020: 69). The atmospheric, affective present of coded understanding is one of the forceful forms through which embodied capital and belonging manifests, woven into a tapestry of privilege. As Wacquant argues, the process of feeling affected and producing affect is a bringing together of history, culture and common assumptions that is co-constituted 'between skilled agent and pregnant world' (Wacquant, 2014: 5). Thinking of place as having an affective history also points towards erasure and the domination of one set of hierarchies over another, expressed both culturally and materially. Within Barangaroo, the mark of erasure manifests as 'urban regeneration', which is little more than the concreting-over of its past as a dockyard – continuing the architectural trend of financial monuments erasing signs of labour (Spencer, 2016) – as well as the continuation of Barangaroo sitting on the unceded lands of the Gadigal people. In other words, atmospheric presence comes with an equal claim to erasure that within any given setting is the formation of absence of 'history'. A claim on this will, as Threadgold (2020) argues, build an understanding of how multiple possible futures are present in any given from of space – locking in gendered constructions of identity and not just belonging within our present, but reinscribed into our pasts and projected into our future. The anywhereness of the International Style that characterizes CBDs like Barangaroo encourages dehistoricization, a glossing-over of the live histories of colonial violence and gender inequality.

Affective atmospheres of glass, reflection and light

Any examination of Barangaroo's atmospheres necessitates our reflecting on the oppressive presence of glass (Figures 11.4 and 11.5). Glass helps to obfuscate the contradictions of Barangaroo, posing as a project to renew public life in an otherwise lifeless space. Such life is window-dressing, a distraction from the dehumanizing influence of capital driving this place making. Baudrillard (1996: 42) describes glass as 'the most effective conceivable material expression of the fundamental ambiguity of "atmosphere", … at once proximity and distance, intimacy and the refusal

Figure 11.4: South towards Crown Casino, along the former dockyard Foreshore

Source: Alex Simpson

Figure 11.5: Reflective properties of glass, Barangaroo

Source: Alex Simpson

of intimacy, communication and non-communication'. So, while glass can be transparent, implying inclusion, it can just as easily exclude, as if it were packaging – we are allowed to see, but we are prevented from feeling, knowing our environment in any lived sense. The atmosphere of glass functions in Barangaroo as an ontological camouflage, not just hiding in plain sight, but making us complicit in the normalization of financialization. Let us not forget, the Barangaroo project is one of private finance's capture of public space. Privatizing a social space, in this way, our exclusion is not immediately perceptible – in no small part thanks to the deployment of enchanting glass architecture. This fallacy, of glass's transparency facilitating belonging, is well observed in Fredric Jameson's (1991) famous description of the Westin Bonaventure Hotel in Los Angeles. Like Jameson's essay, the glass of Barangaroo is a skin that serves to only repel the city outside, affecting a peculiar and placeless dissociation from its neighbourhood. Once again, the most striking feature of Barangaroo is not so much its presence, but its dislocation – both you and this architectural formation could be anywhere. As Jameson (1991: 42) notes, 'you cannot see the hotel itself, but only the distorted images of everything that surrounds it'. Likewise, Barangaroo's enveloping wall-structures generate a mirage of public vitality, with surfaces of financial architecture bouncing off one another like in a hall of mirrors and framed by the glistening waters of the harbour. This functions to mask inner financializing machinations and frames an axis of inclusion and exclusion.

Glass invites us in, but we should be under no illusion that is seeks to share with us what lies beneath. The contradiction of glass is that its very transparency sets up, as Baudrillard (1996: 42) argues, a perceptive but invisible materiality that prevents 'communication from becoming a real opening onto the world'. Like the slippery ephemerality of finance, Barangaroo's glass architecture imposes an asymmetrical power relationship, engendering an atmosphere of ubiquity and distance with belonging marked by an ability to pass through the phantom glass doors of these edifices and step through to the clinical, inner world of financial production. While glass architecture proposes to create a new subjectivity, an altered state of consciousness as the subject dissolved, empathetically, to be at one with the world (Arbab, 2010), it can, as Barangaroo demonstrates, be equally used to de-subjectify. Simpson (2013: 351) refers to architect Reinhold Martin's comparison of glass curtain walls with television, 'a medium to be watched in passing rather than looked at like an artwork'. In a contemporary update, Barangaroo could be said to be swamped by buildings that resemble black mirrors – attention-sapping architectural avatars of the smart phone era (Figure 11.6). This is, once again, to revisit the way atmospheres stand as a mark of erasure as well as unity with, here, glass representing a bland, uniform blankness that erases all senses of temporality from space (Spencer, 2018). Glass becomes an atmospheric point of erasure, removing the 'stains' of labour as well as crudely creating a facsimile of the past, with the weakest

Figure 11.6: Layers of glass, bridge across Sussex Street

Source: Alex Simpson

of gesticulations to the Gadigal people through its name – Barangaroo. The name, like the material, is an object made of glass; an object that has no 'aura' and cleansed from the stain of history (Benjamin, 1999: 734).

To describe glass as a technology that distorts memory and sense making is not new. Douglas Murphy (2012: 7), for example, writes how early glass architecture spectacles, namely Albert Palace and the Crystal Palace at Hyde Park and Sydenham, were always already transient, fragile, weak. They don't decay like a conventional ruin but 'disappear without trace, leaving nothing behind but ephemera'. Such 'abstract ruins' were, thanks in no small part to the spectral qualities of glass, 'inherently anti-monumental' and 'anti-archival' (Murphy, 2012: 79). It is a material that is removed from the past, creating an ever-certain sense of presence, and projects to a future without decay – an architectural equivalent to the 'endless flatlands' of capitalist realism of Mark Fisher (2009) as well as a techno-accelerated future without decay. All of which can be read, like Walby (2015), as gendered relations of bodies, fragility and a desire to overcome our corporeal limitations (Riach and Cutcher, 2014; Simpson, 2019). Rather than woven into financial practices, we now have it written into the atmospheric tapestry of place. The aura-less reflective glass in architecture reinforces a dreamlike sense of unreality and alienation, creating an opaque transparency that only reflects our gaze back to us, unaffected and unmoved (Pallasmaa, 2012). Try as we might, we are unable to see or imagine life behind these walls, reinforcing our ontological sense of disposition and belonging.

Glass is, on the one hand, a cold, sharp and de-sensualizing feature of financial landscapes, but as a material it carries an ocular supremacy that fetishizes light (Pallasmaa, 2012). At night, these edifices dazzle and blind, both figuratively and somewhat physically (Figure 11.7). The fetishization of light in such megaprojects invites a utopian presumption, reassuring the public of, at least literally, a bright future. Light dissolves the facades of these buildings and, pouring through the glass windows, it blinds rather than illuminates (Schmid, 2022). Cut against the black sky, Barangaroo is rendered transparent yet cut by the splashes of neon white that reflect and ripple of the harbour. All shadows are erased. Yet, Barangaroo's surfeit of light deserves deeper contemplation. Pallasmaa (2012) warns of the deflationary consequences of excising dim light and shadows from our world. In other words, 'homogenous bright light paralyses the imagination in the same way that homogenization of space weakens the experience of being, and wipes away the sense of place' (Pallasmaa, 2012: 46). Indeed, the beneficence of light should not be assumed, as Pallasmaa (2012: 49) goes on to remind us. The use of a constantly high level of illumination that leaves no space for mental withdrawal or privacy, itself is a common method of torture. Barangaroo's fetishization of light relegates other senses, other means of knowing and of belonging. Again, drawing on Jameson (1991: 42), the inverted capacity of

Figure 11.7: Barangaroo at night from Darling Harbour

Source: Alex Simpson

light continues to render the outer walls invisible, only a fuzzy distortion of the interior to give a wavering outline of its skeleton. Glass's lack of tactility dampens our ability to *feel* a connection with this place and to more readily comprehend the histories erased by its presence – including structural gender inequalities. Barangaroo seeks to dictate the types of affects generated by its atmospheres, and its use of glass hints at a desire for a seductive but ultimately soulless experience for visitors. Our decaying bodies and imperfect decisions run counter to the structuring logic of finance, and so our belonging to this cathedral to capital can be only superficial.

Understood in this way, the architectural use of glass, as well as the corresponding relation to the 'beams of light', frames the material, representational and affective features of financial architecture – in Barangaroo or elsewhere in the world (Simpson, 2013). Glass is the frame through which the global, financial city is built. Bauhausian ambitions for glass as the material for constructing 'cathedrals of socialism' has, rather than ushering in a 'new unity', become the hallmark of environments built for finance and actively inhospitable to a public (Hatherley, 2011; Atkinson, 2016). Such a contemporary comparison supports Baudrillard's (1996) characterization of our obsession with glass, given its associations with hygiene, as a cultural disavowal of organic functions for a radiant and functional objectivity. If glass rejects history, it also rejects our most human qualities, creating an environment not for community but for an antonymic financialization. Rose describes transparent space as the territory of the oppressors, her experiences of such places resembling that of a trapped bee: 'the language of bumping against invisible barriers, of dead ends, of being jostled and bruised by sharp appraising glances, is a language of a body being defined by powerful others who control the view' (Rose, 1993: 148). The common usage of idioms 'glass ceiling' and 'glass cliff' to describe women's experiences in finance bolsters such an entomological analogy. Glass, coded as hygienic, cold, rational, open, emits what Rose (1993: 7) terms 'masculinist rationality', signalling these institutions and those of them as untainted by the corporealized freight of the past: their thoughts autonomous, context free, objective and universal.

Barangaroo marks a new frontier in Sydney's urban development, one which, as Tafuri (1987: 172) insists, should not be treated as architecture, but sheer instruments of economic policy. Barangaroo is a materialization of the abstract dimensions of financialization, represented by 'extreme isometric spaces' that reflect a prioritization of low-latency movement, and 'enclosed skin volumes' that maximize space to such an extent that they affect a dematerialization, mass and weight decrease while volume and the contour are enhanced. In turn, Barangaroo abstracts the social, dispossessing us all, in a way, and anchoring the structuring logic of financialization in how we physically interact with this environment as well as the affective resonations that reverberate through space.

Conclusion

Focusing on Sydney's Barangaroo development, this chapter has examined how affective atmospheres reveal and give a sense to gendered dynamics of materiality, belonging and presence. The framing of affective atmospheres helps to articulate the multiple ways material and social coding of place reverberates through spatiality, affect and the assembling of bodies, emotion, discourse, materiality and technology (Brand, 2023). Important here is the way these dynamics entwine to give rise to a 'character' of spatial form – one which can be described as an 'atmosphere' that produces gendered frames of inclusion and exclusion. In other words, by creating and arranging material fragments of light, sounds, symbols and texts, atmospheres are 'enhanced', 'transformed', 'intensified', 'shaped' and otherwise impacted upon to create affective qualities of identity, belonging and the sensor force of being *in* or *out* of place (Böhme, 2006; Anderson, 2009). While affective atmospheres have been used to create a new imaginary to shed light on how people, buildings and technologies, including forms of non-human life, come together in a particular geographical setting (Bissell, 2018: 272), this chapter has sought to include the framing of gender within this debate. After all, how we experience the formation of place, as well as its atmospheric flows, will be shaped by – and shape – gendered dispositions. It is in this context that we can see the production and framing of Barangaroo as not just a system of financial place making but one which privileges a masculinist atmosphere that reflects and is entwined with the masculinity of finance as a profession. In other words, if an atmosphere, as Brennan (2004: 1) suggests, 'literally gets into the individual', then the production and encounters of financial architecture and its imprint on urban design – from its exclusivity to the ubiquity of its tough, glass exoskeleton – become a prism that shapes not just our encounters with atmospheres of being in/out of place but the gendered dynamics of identity formation and belonging within.

While, much of the time, atmospheres are almost imperceptible, filtering through the affective layers of material and social organization that give meaning to space, there are moments when atmospheres come to the surface and are more apparent (Anderson, 2009; Young, 2019; Brand, 2023). Traversing the boundaries of place formation, crossing the perceived but sometimes invisible line from one spatial frame to another, can give rise to gendered encounters of belonging that flow between the materiality of place and the bodies that permeate through its domain. It is felt force of relations that holds the affective capacity to bring people together and bring to life, as Sack (1993: 326) emphasizes, concepts of *guest*, *stranger* and *citizen*. Materiality and place making, after all, reaffirm a broader *field of power* to create a material environment that generates a social selective sorting of people and practices within the spatial framework of finance (Threadgold, 2020). These sensory, emotive and embodied affections in response to place create, in turn, an 'energy of feeling' that may arise but dissipate, depending on the context of environmental emersion, but which are encoded by the masculinist disposition contained within finance work and financial place-making projects, such as Barangaroo. It is a spatial and atmosphere formation that reaffirms a felt sense of belonging, or *citizenship*, among its members (Simpson, 2022). The importance of the boundary, here, is to capture those immediate, somewhat imperceptible, impressions that press on our consciousness. A felt force of relations that within the immediacy of the 'first appearance' lay the foundations for the 'later sensations of unreality' (Sechehaye, 1951: 19), whether it is the changing formation and appearance of the swirling of bodies around you, or the towering glass edifices that reflect the brilliant Sydney light to create a gloss and material reverberation of impenetrable smoothness.

Within such moments of encounter exists the affective relations of space, located within the inter-material and inter-social atmospheres that shape our sense of belonging, identity and affinity within any given locale. Here, the materiality of space, specifically the near-ubiquitous presence of glass, lends a cold, steely, de-sensualizing present that becomes a marker of both location and dislocation, as these edifices lose their sense of location by reflecting – or absorbing – all that which surrounds them. The symbolic resonance between financial architecture and the sealed-in, suited bodies that permeate through this space does not need to be spelled out, but what is key is an atmosphere in which these buildings, like the masculinist idealized sense of self, do not decay but, rather, stand impervious before disappearing without trace (Murphy, 2012: 79; Riach and Cutcher, 2014). Barangaroo becomes, in this context, both an atmospheric point of erasure, removing the 'stains' of labour as well as crudely creating a facsimile of the past, and a symbolism of perma-presence which carries an ocular supremacy, stretching forward to claim a future to come. In this way, drawing on Benjamin (1999: 734), Barangaroo carries an 'aura', or atmosphere, that is cleansed from the stains of time and claims a masculinist sense of purpose, presence and belonging.

References

Abraham, N. and Torok, M. (1972) Mourning or Melancholia: Introjection versus incorporation. In: N. Rand (ed) *The Shell and the Kernel*. London: University of Chicago Press, pp 125–138.

Ahmed, S. (2004) The Cultural Politics of Emotion. Edinburgh: Edinburgh University Press.

Anderson, B. (2009) Affective Atmospheres. *Emotion, Space and Society*, 2(2): 77–81.

Arbab, M. (2010) Glass in Architecture. *International Journal of Applied Glass Science*, 1(1): 118–129.

Atkinson, R. (2016) Limited Exposure: Social Concealment, Mobility and Engagement with Public Space by the Super-rich in London. *Environment and Planning A*, 48(7): 1302–1317.

Baker, T. and Ruming, K. (2015) Making 'Global Sydney': Spatial Imaginaries, Worlding and Strategic Plans. *International Journal of Urban and Regional Research*, 39(1): 62–78.

Baudrillard, J. (1996) *The System of Objects*. London: Verso.

Benjamin, W. (1999) *Selected Writings: Volume 2, 1927–1934*. Cambridge, MA: Belknap Press of Harvard University Press.

Bille, M., Bjerregaard, P. and Sørensen, T. F. (2015) Staging Atmospheres: Materiality, Culture and the Texture of the In-between. *Emotion, Space and Society*, 15: 31–38.

Bissell, D. (2018) *Transit Life: How Commuting is Transforming our Cities*. Cambridge, MA: MIT Press.

Böhme, G. (1993) Atmosphere as the Fundamental Concept of a New Aesthetics. *Thesis Eleven*, 36(1): 113–126.

Böhme, G. (2006) Atmosphere as the Subject Matter of Architecture. In: P. Ursprung (ed) *Herzog and Meuron: Natural History*. London: Lars Müller Publishers, pp 398–407.

Bourdieu, P. (2010) *Distinction: A Social Critique of the Judgement of Taste*. London: Routledge.

Brand, A. (2023) *Touching Architecture: Affective Atmospheres and Embodied Encounters*. Milton: Taylor and Francis Group.

Brennan, T. (2004) *The Transmission of Affect*. Ithaca, NY: Cornell University Press.

Burns, S. (2019) Arrivals and Departures: Navigating an Emotional Landscape of Belonging and Displacement at Barangaroo in Sydney, Australia. In: C. Courage and A. McKeown (eds) *Creative Placemaking: Research, Theory and Practice*. London: Routledge, pp 56–68.

Edvardsson, D., Rasmussen, B. H. and Riessman, C. K. (2003) Ward Atmospheres of Horror and Healing: A Comparative Analysis of Narrative. *Health*, 7(4): 377–396.

Fisher, M. (2009) *Capitalist Realism: Is There No Alternative?* Winchester: O Books.

Frosh, S. (2012) Hauntings: Psychoanalysis and Ghosly Transmission. *American Imago*, 69(2): 241–264.

Gregg, M. and Seigworth, G. J. (2010) An Inventory of Shimmers. In: M. Gregg and G. J. Seigworth (eds) *The Affect Theory Reader*. Durham, NC: Duke University Press, pp 1–25.

Griffero, T. (2014) Atmospheres and Lived Space. *Studia Phaenomenologica*, 14: 29–51.

Harris, M. (2018) Barangaroo: Machievellian Megaproject or Erosion of Intent. In: K. Ruming (ed) *Urban Regeneration in Australia: Policies, Processes and Projects of Contemporary Urban Change*. London: Routledge, pp 111–134.

Hatherley, O. (2011) It Is Old Corruption in Braced Glass. *Building Design*, 14, https://www.bdonline.co.uk/city-of-london-old-corruption-in-bra ced-glass/5028233.article.

Hayden, D. (1977) Skyscraper Seduction, Skyscraper Rape. *Heresies*, 2: 108.

Jameson, F. (1991) *Postmodernism, or, The Cultural Logic of Late Capitalism*. Durham, NC: Duke University Press.

Kern, L. (2021) *Feminist City: Claiming Space in a Man-made World*. London: Verso.

Lamont, M. and Molnar, V. (2002) The Study of Boundaries in the Social Sciences. *Annual Review of Sociology*, 28: 167–195.

Lefebvre, H. (1991) *The Production of Space*. Oxford: Blackwell.

McCormack, D. P. (2008) Engineering Affective Atmospheres on the Moving Geographies of the 1897 Andrée Expedition. *Cultural Geographies*, 15(4): 413–430.

McDowell, L. (1997) *Capital Culture: Gender at Work in the City*. Oxford: Blackwell Publishers.

Murphy, D. (2012) *The Architecture of Failure*. Winchester: John Hunt Publishing.

Noble, G. (2013) 'It is Home but it is not Home': Habitus, *Field and the Migrant Journal of Sociology*, 49(2–3): 341–356.

NSW Government (2005) *East Darling Harbour, Sydney: Urban Design Competition Brief.*

OMFIF (2023) *Gender Balance Index 2023.* Available at: www.omfif.org/gbi2023/ (accessed 26 February 2024).

Pallasmaa, J. (2012) *The Eyes of the Skin: Architecture and the Senses.* Hoboken, NJ: John Wiley and Sons.

Rancière, J. (2007) *The Future of the Image.* London: Verso.

Riach, K. and Cutcher, L. (2014) Built to Last: Aging, Class and the Masculine Body in a UK Hedge Fund. *Work, Employment and Society,* 0(0): 1–17.

Rose, G. (1993) *Feminism and Geography: The Limits of Geographical Knowledge.* Cambridge: Polity.

Sack, R. (1993) The Power of Place and Space. *Geographical Review,* 83(3): 326–329.

Sassen, S. (2005) The Embeddedness of Electronic Markets: The Case of Global Capital Markets. In: K. Cetina and A. Preda (eds) *The Sociology of Financial Markets.* Oxford: Oxford University Press, pp 17–37.

Sassen, S. (2019) *Cities in a World Economy.* Thousand Oaks, CA: SAGE Publications, Inc.

Schmid, M. (2022) Architecture of Apocalypse, City of Lights. In: M. Schmid (ed) *Intermedial Dialogues: The French New Wave and the Other Arts.* Edinburgh: Edinburgh University Press, pp 127–162.

Sechehaye, M. (1951) *Autobiography of a Schizophrenic Girl: Reality Lost and Gained, with Analytic Interpretation.* New York: Grune and Stratton.

Simpson, A. (2019) Establishing a Disciplining Financial Disposition in the City of London: Resilience, Speed and Intelligence. *Sociology,* 53(6): 1061–1076.

Simpson, A. (2022) A Relational Approach to the Ethnographic Study of Power in the Context of the City of London. *Ethnography.* DOI: 10.1177/14661381221145816.

Simpson, T. (2013) Scintillant Cities: Glass Architecture, Finance Capital, and the Fictions of Macau's Enclave Urbanism. *Theory, Culture and Society,* 30(7–8): 343–371.

Spencer, D. (2016) *The Architecture of Neoliberalism: How Contemporary Architecture Became an Instrument of Control and Compliance.* London: Bloomsbury Publishing Plc.

Spencer, D. (2018) Agency and Artifice in the Environment of Neoliberalism. In: E. Wall and T. Waterman (eds) *Landscape and Agency: Landscape and Agency Critical Essays.* London: Routledge.

Stewart, K. (2007) *Ordinary Affects.* Durham: NC: Duke University Press.

Sudjic, D. (2005) *The Edifice Complex: How the Rich and Powerful, and Their Architects, Shape the World.* London: Penguin.

Tafuri, M. (1987) *The Sphere and the Labyrinth: Avant-gardes and Architecture from the Piranesi to the 1970s.* Cambridge, MA: MIT Press.

Threadgold, S. (2020) *Bourdieu and Affect: Towards a Theory of Affective Affinities.* Bristol: Bristol University Press.

Thrift, N. (2004) Intensities of Feeling: Towards a Apatial Geography of Affect. *Geografisko Annaler Series B*, 86: 57–78.

Urry, J. (2007) *Mobilities.* Cambridge: Polity.

Wacquant, L. (2014) Homines in Extremis: What Fighting Scholars Teach Us about Habitus. *Body and Society*, 20(2): 3–17.

Walby, S. (2015) *Crisis.* London: Polity.

Weisman, L. K. (1981) Women's Environmental Rights: A Manifesto. *Heresies*, 11: 6.

Young, A. (2019) Japanese Atmospheres of Criminal Justice. *The British Journal of Criminology*, 59(4): 765–779.

Young, A. (2021) The Limits of the City: Atmospheres of Lockdown. *British Journal of Criminology*, 61(4): 985–1004.

12

Liminality and Affect: Knowing and Belonging among Unscripted Bodies

Nyk Robertson

Introduction

The imagining of possibilities for bodies and relationships is influenced by the scripts presented to us as we navigate our own identity development. Through familial and social structures, we observe scripts that inform the ways we connect to our own bodies, and interact with other bodies, as well as spaces in which we move. These scripts are often prescribed and presented in a heteronormative and cisnormative way. These scripts are linked to the colonization of bodies and to racism and the way it informs validity and the valuing of certain prescribed scripts over others. This negotiation of scripts dictates what is acceptable and what is considered deviant. For bodies that do not authentically follow these prescribed scripts, finding possibilities of being outside of them can become necessary to be 'known' and find a sense of belonging. I argue here that these unscripted bodies must move into liminal spaces to detach themselves from prescribed scripts. This detachment can have a cost and often comes with loss. However, within liminal spaces, and within movement of spaces and identity, a freedom and hope of possibilities also exists. Queer bodies have the potential to rewrite scripts and create 'possibility scripts' that can enable space for other queer bodies, and queer connections within and between bodies and space. Intersex bodies offer a layer of complexity to how bodies exist and interact with one another and the spaces they inhabit. The focus of this chapter will be queer bodies as they relate to sexuality and gender identity. However, it is important to recognize the existence of intersex bodies and the impact these bodies could have on affect and possibility scripts.

The aim of this chapter, therefore, is to explore the ways in which affect and queer bodies are 'in conversation' with one another and to consider the ways in which assessment of this conversation can bring about spaces of belonging and community for queer bodies. Through this exploration, I hope to initiate research that affects the ways space is intentionally created so that possibility scripts are made more accessible to queer communities in the future. First, this chapter will define affect and queerness and the relationship between the two. Next, I will examine prescribed scripts presented to us, how they are presented and the ways in which affect is connected to these prescribed scripts. After assessing prescribed scripts, the chapter investigates liminal spaces as well as the power and loss that can come with existing in liminal spaces. Lastly, I discuss the ability to create possibility scripts and how the ability to be known and know others can create community and a sense of belonging.

Affect and queer bodies

Affect exists beyond emotions and feelings. Affect, like many queer bodies, 'is beyond cognition and always interpersonal' (Pile, 2010: 8). In the ways that bodies are understood through their relationship with others and space, affect is understood in ways that are connected to bodies and space and is not easily defined. Affect is complex and gives depth to our existence through relations with others and ourselves (Tomkins, 1963). Both affect and queerness are difficult to define clearly. There are complexities, layers and intersections that relate to fluidity and their inability to be defined neatly. Affect, and queer bodies, hold possibilities for social change and understanding connections beyond definition. Affect's difference from social structures gives it the capacity to restructure social meaning (Massumi, 2002; Sedgwick, 2006). While bodies are introduced to prescribed scripts though social structures, affect in queered spaces can expand possibilities.

Scripted bodies

Through socialization, our bodies are given a script of what is possible, how we are to perform our identities and how they can connect (Butler, 1997; Cavarero, 2010; Ahmed, 2015). This normative script is interrelated to family, culture, social norms, heteronormativity and cisnormativity. Compulsory heterosexuality shapes bodies in relation to other bodies and the ideal relationships formed in their interaction (Ahmed, 2015). The ideal relationship is linked to the ideal family, including having access to a legal marriage, being able to create children and increase the number of members of the family unit. These social structures are presented when attending family functions, viewing your own family dynamics growing up, seeing families portrayed in media and even having teachers and other authoritative figures refer to specific kinds of families as being valid. For example, hearing a teacher tell the class to have their mums and dads attend parent–teacher conferences reiterates this script. The common place of compulsory heterosexuality is what makes it affective. The way families are formed, celebrated and bounded is linked to this compulsory heterosexuality (Ahmed, 2015). Attending weddings, baby showers and other family gatherings assists in impressing this script onto us as we develop and increase our understanding of familial and intimate relationships.

Conversely, the policing of other kinds of relationships impresses upon what is valid and legitimate. 'Normative culture involves the differentiation between legitimate and illegitimate ways of living' (Ahmed, 2015: 149). Through laws and culture, behaviour and people become defined as deviant or even invisible. In many countries, marriage is not legal between bodies that are not sexed in ways that are defined as opposite to one another. The ways in which we view stepparents, adopted parents or children being raised by their grandparents are other examples of legitimizing ways of living. Having language and definitions to delineate types of caregivers for children creates a hierarchy of legitimacy and ideal familial relationships. By delegitimizing certain ways of living, compulsory heterosexuality is represented as the script one needs to follow to fully participate in a legitimate romantic relationship. A characteristic of this relationship is that the bodies involved in this relationship must be defined in certain ways individually and in relationship to each other. These scripts can look different based on culture, geographical location, socio-economic status and other spaces a person occupies. While prescribed scripts are clearly defined by groups a person is part of or raised with, these scripts are not hegemonic across all experiences and identities.

Affect and social meaning

Our bodily scripts are influenced by those closest to us. Our families affect our understanding and shaping of ourselves and the world (Cavarero, 2010). Through repetition and performativity, we make meaning of gender and passing bodies (Butler, 1997). Our understanding of women and men, femininity and masculinity, as well as how these bodies interact with each other, is influenced by how we see these things performed. In order to understand how scripts are reproduced, we must understand how affect interacts with performativity and meaning making in the same way that language and signifiers interact.

While bodies have scripts impressed upon them in relation to themselves and their interaction with others, these bodies are also affected by the spaces which they inhabit. Spaces are designed in ways that extend validity and understanding of bodies. Gendered spaces and gendered policing reaffirm binary genders and leave these defined identities unquestioned (Browne et al, 2020). The relation of bodies and interaction between them also affects how they enter social spaces (Ahmed, 2015). For instance, physical displays of intimacy between two bodies that are perceived as male bodies may create feelings of uneasiness in heteronormative social spaces. These displays of intimacy can also be linked to the way in which emotions and feelings are displayed or accepted. How bodies of certain perceived genders are permitted to interact in social spaces can be linked to how emotions and feelings are linked to displays of these emotions. 'Emotions [are] framed and circumscribed [by] sexed and gendered experiences of place and spaces' (Pile, 2010: 7). Certain emotions are linked to femininity and masculinity, which also prescribe which bodies can display them. In this way, bodies are defined not only by their physical characteristics but by how those bodies interact with one another, and in what spaces.

If a body fails to fit into the prescribed script, they often become invalid, othered, defined as a failure or unintelligible. Identities become othered by their failure to be the ideal (Ahmed, 2015). This rejection of the ideal also suggests that there is an authentic, natural body from which to deviate (Blackman and Venn, 2010). While this social contract of what is authentic and natural can influence how bodies present themselves, or are understood, there is an option to rewrite the social script in a way that creates space for other bodies and relationships.

Rewriting the script

Transgender bodies do not follow traditional scripts of gendered bodies. These bodies move outside of understood borders within sex and gender. In doing so, transgender bodies reveal

> [a]s fraudulent the accepted version of the relations between sex and gender in which sex is thought to be the natural cause of gender. The [transgender] subject's role is that of a debunker, unveiling this representation of sex to be just that, a representation or simulation, not the natural cause or ground of gender at all, but its projection. (Prosser, 1995: 483)

By existing in bodies that illustrate the performativity of the relationship between sex and gender, trans bodies bring into perspective the scripts that exist for all gendered bodies. Power lies within this potential of choosing to be, or being placed, outside of prescribed scripts. Systems 'seek to normalize, categorize and fix the proper relations of objects, this makes [queer] difficult to define, categorize and most importantly control' (Browne, 2006: 889). 'Queer feelings are "affected" by the repetition of the scripts that they fail to reproduce, and this "affect" is also a sign of what queer can do, of how it can work by working on the (hetero)normative' (Ahmed, 2015: 155). Queering scripts placed on bodies, relationships between bodies and the spaces they exist within constructs a possibility of hope. The hope of queer includes the possibilities of social forms not constrained by heteronormativity, and the abilities to affect and change social spaces (Ahmed, 2015). Queer bodies illustrate possibilities and allow other bodies to envision an existence outside of heteronormative scripts of being.

Rewriting prescribed scripts not only requires stepping outside of what is known and performed by our social communities, but also requires a reorientation into new boundaries and connections, including familial relationships (Aramburu Alegría, 2018; Kelley, 2020). In order to be understood by others, queer bodies can either inauthentically follow scripts prescribed to them or reorient those with whom they wish to be intelligible. There is a messiness to affective reorientations and to navigating social expectations (Perger, 2022). Possibilities of being and being known are connected to reorienting not only one's body but also how one's body is perceived by others and the relationship with other bodies. 'Bodies are always thoroughly entangled processes, and importantly defined by their capacities to affect and be affected' (Blackman and Venn, 2010: 9). In order to create possibility scripts, bodies must be in relation with other bodies and with the spaces they occupy.

Familial relationships often offer prescribed scripts as well. These scripts can include having a daughter, raising a son, or a parent imagining themselves walking their daughter down the aisle. Bodies, and their romantic relationship with other bodies, are directly linked to familial scripts. 'Kinship "matters" in the way that bodies "matter": it may be produced or constructed but is no less urgent or tangible for that' (Freeman, 2007: 298). Rewriting scripts may require the reforming of kinship, or even finding new kinship. Chosen families may take the place of the kinship connection as bodies step outside of compulsive heterosexual familial scripts. 'The gap between the script and the body, including the bodily form of "the family", may involve discomfort and hence may "rework" the script' (Ahmed, 2015: 152). This movement can mean the loss of family connections as well as other material losses.

Due to this fear of loss, some queer bodies may choose not to live authentically. 'To follow a different path would be to not only compromise your own happiness but the happiness of others' (Schmitz and Ahmed, 2014: 113). However, upholding traditional family structures at the cost of their transgender identities has brought about shame, guilt and feelings of responsibility (Perger, 2022). In order to create new scripts, and new space, for bodies, loss is inevitable.

Loss

There is a noticeable cost to not following the defined script for our bodies. However, following the script is not always a choice. Intersex bodies exist physically within a space that is not situated within the gender binary. Even with bodies that may fit into the defined gender binary, if these bodies are trans bodies, there are still barriers to acceptance:

> Seeking acceptance within the system of 'normal' and denying our transsexual status is an acquiescence to the prevailing binary gender paradigm that will never let us fit in and will never accept us as equal members of society. Our transsexual status will always be used to threaten and shame us. (Green, 2016: 123)

Shame affects the bodies that hold it within them. There is a relationship between bodies, affect and trauma. This trauma lies outside of a personal experience (Blackman and Venn, 2010). Bodies carry this trauma in forms that show up internally, but also externally in connection to other bodies, and in social spaces that amplify that trauma. The effect of loving bodies that are not supposed to be lovable can bring shame and melancholia (Braidotti, 2002; Butler, 1997). The cost of living authentically outside of prescribed scripts can include being affected by shame, trauma and absence of belonging.

Beyond the cost to bodies, there are also social and material costs to living outside of what is defined as legitimate bodies and relationships. Having relationships not recognized by others comes with social and psychological costs (Ahmed, 2015). In countries where marriages between certain bodies are prohibited, there is a cost of income, access to legal rights afforded to married couples and more. Loss is also tied to systems and the ways in which these systems are set up to only recognize bodies within prescribed scripts. Doan (2010) names these experiences a tyranny of gender and argues that those who transgress gender norms are subject to this tyranny. Making a conscious decision to live outside of intelligible, accepted bodies and relationships is not a privilege everyone can afford. Some may not have access to enough privilege, power or capital to risk being illegitimate (Ahmed, 2015). Being illegitimate means losing access to spaces, material goods and even familial support. Due to these material and familial losses, some bodies may participate in traditional kinship structures not only as a strategic decision to minimize loss but also out of necessity for safety and survival (Ahmed, 2015). It is important to acknowledge and honour the risk of unscripted bodies living authentically.

Liminality and changing the script

Bodies can find themselves in liminal spaces by choice, or by being placed there due to their illegitimacy. 'There are those who choose a permanent state of liminality, and others who are liminal by circumstance, condition, or social definition. The liminal persona comes to be in two ways, the voluntary and involuntary' (Carson, 2016: 10). Placing oneself outside of understood borders, or being placed there by others, results in finding yourself among other liminal bodies. These bodies may remain in that liminal space indefinitely, or may move into other, new, bordered spaces. 'The dual aspect of liminality as both a desired enduring site of being and a finite process of becoming neatly captures the bind which many trans-individuals appear to be caught in with regards to social and political recognition' (McQueen, 2011: 7). Without defined identities, the access to political and social recognition is diminished, which also can lead to a lack of access to systems such as healthcare, housing and employment. The ability to be affected and affect others depends on the intelligibility between bodies.

Liminality is a fluid state of understanding and being (Meyer and Land, 2005). Liminality allows movement between defined borders of identity. This fluidity allows bodies to explore identities outside of prescribed scripts of gendered bodies. 'Someone whose personhood is liminal lives beyond the pale of society, or structure ... [liminality is] an open-ended way of life qualified by sets of cultural demands, ethical systems, and processes that are irreconcilable ... outsiderhood and marginality defy reincorporation' (Joseph, 2011: 140). In order to be in a fluid state of movement, one must be outside of systems of recognition that are linked to prescribed scripts.

While this liminal state comes with loss, as described earlier, existing in liminal spaces also comes with spaces of freedom. 'Freedom results from a rejection of those rules and norms that have structured social action prior to the liminal situation' (Yang, 2000: 383). By breaking free of prescribed scripts, bodies are able to move into spaces otherwise denied to them. Bodies can exist differently and can affect other bodies differently. This movement between states and bodies is necessary for change (Massumi, 2002). These changes include a change in possible scripts, possible states of being and possible relationships between bodies and spaces.

When considering affect, scripts and liminality, it is important to consider Bandura's (2012) three affective components: hope, optimism and resilience. Movement can exist between trauma and hope as bodies produce new ways of being. Bodily affectivity is organized around 'an ethics of hope and optimism for change' (Blackman and Venn, 2010: 14). Changing what is possible can bring hope to bodies holding shame from their existence outside of prescribed scripts. These experiences with change and trauma can build resilience within queer bodies. Resilience is linked to the ability to cope with liminality and persist within uncertainty (Rattray, 2016). This resiliency allows bodies to develop new scripts and spaces within which these scripts can transpire.

Queer theorists have spent considerable time exploring movement, 'especially when it is movement against, beyond, or away from rules and regulations, norms and conventions, borders and limits' (Epps, 2001: 413). This movement outside of borders, and within liminal spaces, is directly linked to bodies' affects upon one another, and in relation to social spaces.

Liminality, and ambiguity, allow certain bodies to move more freely between bordered spaces. Bodies that can move with ease may also more easily shape and be shaped by the sign queer (Martin, 1996). This capability to move more freely values certain bodies over others. Bodies that are less intelligible may risk loss within clearly defined spaces governed by prescribed scripts, while finding more freedom and acceptance in spaces built for queer bodies.

Within queer spaces, bodies can also be seen as queer without being completely intelligible. The interaction and connection between queer bodies may be read as queerness without always being clearly defined as intelligible or existing in a precise, fixed, understood identity. The fluidity of queerness and identities allows for a movement within queerness that can be authentic and valid without being particularly intelligible. This can create tension between existing within a possibility script and navigating the fluidity of identity. 'There is a risk of solidifying, homogenizing, and "de-queering" through the act of naming' (Browne, 2006: 888). This complicates the ability to know and be known in nuanced and intricate ways.

Knowing and being known

In a 2020 study of 14 non-binary adolescents, participants explained that they did not know there were options outside of the binary, which made it difficult to understand their own feelings about their gender and their bodies. This created confusion that left them questioning whether their identities were real or valid (Johnson et al, 2020). The lack of possibility scripts affected these participants' ability to see themselves. Even when possibility scripts were available, these scripts were prescriptive in what experiences of transness were valid. Participants also reported feeling not connected to a specific experience of gender dysphoria, or a connection to an understood gender made them feel inauthentic or not trans enough (Johnson et al, 2020). Without representation of what is possible, non-binary youth lacked understanding of what was attainable.

Queer experiences have the ability to create possibility scripts while also leaning into the discomfort of the unknown. Queer feelings embrace discomfort in the lack of scripts available for living and loving, but excitement in what living outside of these scripts can mean (Ahmed, 2015). Creating possibility scripts requires movement into liminal spaces not yet defined by prescribed scripts inherited from familial relationships. Creating kinship can be difficult and thrilling when there is an absence of possibility models (Weston, 1997). In order to form kinship, it is necessary to create possibility scripts that are flexible enough to move within liminal spaces without being confined by prescribed borders. Identity making is a 'dynamic process of perpetual resurfacing' that involves impressions of yourself in relation to others (Ahmed, 2015: 160). The process of resurfacing is informed by the possibility scripts that are visible to bodies and in relation to other queer bodies.

To be known, one must be intelligible to others. 'One "exists" not only by virtue of being recognized, but, in a prior sense, by being recognizable' (Butler, 1997: 5). In order to be recognized, bodies must be able to be interpreted in relation to known prescribed scripts. It is difficult to see bodies that exist in liminal spaces that do not violate their sense of non-identification with existing gender and sexual norms. To a certain degree, these bodies become invisible (McQueen, 2011). If bodies cannot be defined by how they affect and are affected by other bodies, the existence of these bodies is itself in question (Butler, 1986). The category of queer is produced in relation to histories that render it an illustration of failed being or non-being (Ahmed, 2015). Without possibility scripts, and without interacting with other bodies who also fall outside of prescribed scripts, it is difficult not only to be known but to know oneself as real and authentic.

Finding common language, and mutual understanding of meaning, can help to define identities and experiences. However, in defining something, borders are put around that identity in a way that excludes bodies and identities that fall outside of the understood meaning of language. Similarly, defining emotions can take away their ability to exist outside of what is understood. Language of emotion and trying to define experiences through language can make them less authentic (Thrift, 2004). In this way, creating language for queer identities can also place them into a space that is less authentic. Conversely, queer bodies affecting one another, and queer spaces offering an experience of identity without defining it, can move queer bodies closer to knowing and being known.

Unscripted bodies

Bodies are understood in how they interact with the world around them. As we form our identities and boundaries, we are influenced by what surrounds us. However, when our bodies feel differently in themselves, and in relation to others and place, there is a disconnect between what is understood to be possible and what feels authentic.

Formation of intelligible bodies

Our bodies have social norms, familial structures and societal boundaries impressed upon them by the world around them. Bodies become shaped by these prescribed scripts in a way that creates something altered and ingenuine. 'How bodies work and are worked upon shapes the surfaces of bodies … through repeating some gestures and not others, or through being oriented in some directions and not others, bodies become contorted' (Ahmed,

2015: 145). By contorting the body into a prescribed script that is intelligible, the body becomes a version of itself. This repeated working and reworking of the body is done constantly and with force (Ahmed, 2015). Bodies find themselves in space and in relation to others continuously. To be seen, and understood, these bodies must be recognizable. However, this continual contorting of the body can do harm to the body itself. Experiences of interpellation repeated over and over can cause bodily injury (Ahmed, 2015).

Prescribed scripts define bodies' ability to connect with, and be in relation to, other bodies. These scripts affect what bodies can do, and with whom (Ahmed, 2015). In understanding the parameters of the ways bodies are permitted to interact with one another, these prescribed scripts govern what are legitimate interactions. These parameters are set through what is modelled as possible, but also by the way social spaces perpetuate prescribed scripts of interaction.

The surface of spaces is already impressed upon by previous heteronormative bodies, which leaves queer bodies feeling uncomfortable in these spaces because they do not fit what has already been shaped (Ahmed, 2015). Interactions between multiple bodies, or between body and space, are prescribed by what has already shaped those spaces and defined valid interactions between bodies and spaces. This shaping of space, once again, requires contorting of bodies that do not authentically fit into these spaces. The reshaping of bodies to fit space creates a tension between individual identity and community.

Comfort is dependent not only on the body but on how it interacts with the external. Comfort can be between a body and object, or multiple bodies, but is dependent on the fit between the body and the external (Ahmed, 2015). If the external follows prescribed scripts, bodies that do not fit within these prescribed scripts cannot find comfort. 'To feel uncomfortable is precisely to be affected by that which persists in the shaping of bodies and lives' (Ahmed, 2015: 155). Spaces that are not built for bodies outside of prescribed scripts are spaces that attempt to shape bodies to fit, as opposed to bodies having the ability to shape those already constructed spaces of clearly defined heteronormativity.

The lack of comfort requires the body to be hyper aware of itself, in the ways it is uncomfortable and in the ways it does not connect fully with the external authentically. This discomfort causes a feeling of disorientation. 'One's body feels out of place, awkward, unsettled ... acute awareness of the surface of one's body, which appears *as* surface, when one cannot inhabit the social skin, which is shaped by some bodies, and not others' (Ahmed, 2015: 148). This social skin, as Ahmed refers to it, serves as a barrier that keeps scripted bodies in and unscripted bodies out. Feelings of awkwardness and discomfort add to the shame and isolation queer bodies feel as they lack connection to scripted bodies and to place.

Space and belonging

While prescriptively scripted spaces can impress discomfort upon bodies that do not fit within these spaces, queer bodies can create spaces that expand into possibility scripts allowing for bodies to interact with the external in an authentic, comfortable way. These spaces include homes of queer chosen families, queer bars and clubs, activist spaces, identity centres, among other spaces. In this way, interpersonal relationships and interactions connect to the larger feeling of community.

Within queer spaces, bodies have the ability to affect and be affected in ways that lie outside of prescribed scripts. Through rituals of bodily call and response, cultural belonging can be found (Freeman, 2007). These queer and queered spaces create external objects and bodies that interact in a way that can provide comfort and belonging. Spaces are claimed through witnessing of others and the ability to inhabit social spaces in a way that forms belonging (Ahmed, 2015). In this way, queer spaces are crucial in providing places for belonging and community.

As Ahmed (2015: 165) describes it: 'Queer bodies "gather" in spaces, through the pleasure of opening up to other bodies … The hope of queer politics is that bringing us closer to others, from whom we have been barred, might also bring us to different way of living with others'.

Queer gatherings create possibility scripts that increase the potential for belonging, as well as for understanding of one's own identity and body and how it relates to others and the spaces they inhabit. While queer spaces can be a place for recognition and belonging, queer spaces are not always comfortable for all queer folx. The intersection of identities and representation often influence how comfortable certain people are in queer spaces. Many queer bars are predominantly white. Some have predominantly cisgender patrons. Classism, ableism and ageism are also present in some queer spaces. Many spaces defined as 'gayboyrhoods' are occupied with cisgender white gay men and are inaccessible to people of colour and not affordable for much of the queer community (Doan, 2010). Additionally, some queer spaces may even cause some to worry about not being queer enough or the right kind of queer (Ahmed, 2015). With freedom from prescribed scripts, queer spaces can sometimes be sites of policing of identities and bodies in another way. The lack of intentional intersectionality, and the reproduction of respectability politics, creates barriers to belonging.

Bodies that do not follow prescribed scripts lack the ability to find stability within their identities and existence. These bodies are continuously navigating how they are existing within spaces, how they are relating to other bodies and how they are being seen and understood or misunderstood. This continual movement necessary to navigate spaces places these bodies in a space of liminality that does not allow for permanence. These bodies are continually renegotiating their authentic and perceived identities and the comfort or discomfort available between themselves and other bodies and spaces.

> Subjective transnationalism also reflects the experience of feeling 'at home' in more than one geographic location, where identity construction is deterritorialized as part of shifting race–ethnic boundaries or gendered transitions in a globalizing world. Conversely, subjective transnationalism includes feelings that one is neither from 'here' nor from 'there' not at home anywhere. (Segura and Zavella, 2008: 540)

The lack of home referenced by Segura connects to spaces and how these spaces impress upon bodies. As bodies that fall outside of prescribed scripts enter rooms, these bodies are continually navigating their own identities as well as the identities and interactions with other bodies in those spaces. Because of this continuous navigating, a permanent space, or home, cannot be realized. Despite the potential for queer spaces and gatherings to create a sense of belonging, it is almost impossible to only exist within those queer spaces, specifically if these bodies are entering into spaces for work, familial relationships, medical care and other systems that fall outside of queer spaces and gathering.

It is important to note that this search for home can disenfranchise bodies that do not fit within prescribed scripts. By placing oneself within a prescribed script, even one that is different from a prescribed script given to a body in childhood, the visibility of unscripted bodies becomes less prominent. 'Home is what the person living in the margins cannot want … the journey home for the transsexual may come at the expense of a recognition that others are permanently dislocated' (Halberstam, 1998: 171). A body placed into an understood prescribed script makes unscripted bodies less visible and decreases the spaces in which queer bodies can find belonging. By finding stability within a prescribed script of gendered bodies, these bodies further push unscripted bodies into the margins. The act of finding a home can, then, displace unscripted bodies into an increased state of homelessness, which could be described as a gentrification of gender. As bodies place themselves within prescribed scripts of gender, those who cannot exist within prescribed scripted bodies are forced further into the margins (Robertson, 2018). While unscripted bodies may not find a permanent state of home, queer spaces construct liminal spaces of belonging, or being known and knowing.

Conclusion

Unscripted bodies move in and out of spaces where they are known and unknown, as well as intelligible and illegitimate. These bodies can be seen to exist in liminal spaces where they relate to other bodies, and to spaces of comfort and discomfort. Queer bodies are positioned within a possibility of hope connected to freedom and resilience. These bodies, and queer spaces, help to create possibility scripts. By being outside of prescribed scripts of gender, these bodies navigate shame, lack of acceptance and loss of access. However, they also can live more authentically, find belonging, be known and better know each other.

Throughout this chapter, I have presented an argument that better understanding the relationship between affect and queerness has the potential to create space for unscripted bodies and build a sense of belonging among queer bodies. As was presented in this chapter, unscripted bodies often struggle to find community as they develop their identities. This lack of community comes from a lack of possibility scripts, which can cause an inability to feel valid and real. Better understanding of how affect contributes to feelings of belonging, or lack of, allows more inclusive decisions to be made around the creation of space and access to possibility scripts.

The hope for this chapter is to establish a foundation for further research that can aid in creating space and access to possibility scripts in an effort to enhance the lived experiences of queer bodies. Further research could include how affect and intersex bodies are related, as well as the ways in which biological sex and systems that define biological sex within the binary affect experiences of queer bodies. The materialistic effect and access to resources that are tied to these systems directly affect the lives of queer bodies. Another area for further research is access to queer spaces and representation within those queer spaces. While this chapter has explored the possibilities created through queer spaces, it is important to recognize that these spaces are not accessible to all queer bodies and often lack representation of many identities within the queer community, including people of colour and people with disabilities. For affect and space to truly be a place for freedom and possibilities, they need to be accessible to all bodies within the queer communities. To further investigate accessibility and representation, it is necessary to explore the ways in which colonialism and white supremacy inform hegemonic prescribed scripts. The aim of this chapter has been to start the conversation while recognizing that there needs to be further research.

References

Ahmed, S. (2015) Queer Feelings. In: *The Cultural Politics of Emotion*. New York: Routledge, Taylor and Francis Group, pp 148–167.

Aramburu Alegría, C. (2018) Supporting Families of Transgender Children/Youth: Parents Speak on Their Experiences, Identity, and Views. *International Journal of Transgenderism*, 19(2): 132–143. https://doi.org/10.1080/15532739.2018.1450798

Bandura, A. (2012) *Self-efficacy: The Exercise of Control*. New York: W. H. Freeman.

Blackman, L. and Venn, C. (2010) Affect. *Body and Society*, 16(1): 7–28. https://doi.org/10.1177/1357034x09354769

Braidotti, R. (2002) The Politics of Ontological Difference. In: T. Brennan (ed) *In Between Feminism and Psychoanalysis*, London: Routledge, pp 89–105.

Browne, K. (2006) Challenging Queer Geographies. *Antipode*, 38(5): 885–893. https://doi.org/10.1111/j.1467-8330.2006.00483.x

Browne, K., Lim, J., Hall, J. and McGlynn, N. (2020) Sexual(ities that) Progress: Introduction. *Environment and Planning C: Politics and Space*, 39(1): 3–10. https://doi.org/10.1177/2399654420954213

Butler, J. (1997) *Excitable Speech: A Politics of the Performative*. Abingdon: Routledge.

Butler, J. (1986) Variations on Sex and Gender: Beauvoir, Wittig, and Foucault. *Praxis International*, 5(4): 505–516.

Carson, T. L. (2016) *Liminal Reality and Transformational Power*. Cambridge: The Lutterworth Press.

Cavarero, A. (2010) *Relating Narratives, Storytelling and Selfhood*. London: Routledge.

Doan, L. (2010) The Tyranny of Gendered Spaces – Reflections from beyond the Gender Dichotomy. *Gender, Place and Culture*, 17(5): 635–654. https://doi.org/10.1080/0966369x.2010.503121

Epps, B. (2001) The Fetish of Fluidity. In: T. Dean and C. Lane (eds) *Homosexuality and Psychoanalysis*, Chicago: Chicago University Press, pp 412–431.

Freeman, E. (2007) Queer Belongings: Kinship Theory and Queer Theory. In: G. E. Haggerty and M. McGarry (eds) *A Companion to Lesbian, Gay, Bisexual, Transgender, and Queer Studies,* Chichester: Wiley & Sons, pp 295–314.

Green, J. (2016) Look! No, Don't! The Visibility Dilemma for Transsexual Men. In: K. More and S. Whittle (ed) *Reclaiming Genders: Transsexual Grammars at the Fin de Siècle*, New York: Bloomsbury, pp 117–131.

Halberstam, J. (1998) *Female Masculinity*. Durham, NC: Duke University Press.

Johnson, K. C., LeBlanc, A. J., Deardorff, J. and Bockting, W. O. (2020) Invalidation Experiences among Non-binary Adolescents. *The Journal of Sex Research*, 57(2): 222–233. https://doi.org/10.1080/00224 499.2019.1608422

Joseph, M. (2011) *Keywords for Children's Literature*. Albany, NY: New York University Press.

Kelley, A. D. (2020) Cisnormative Empathy: A Critical Examination of Love, Support, and Compassion for Transgender People by Their Loved Ones. *Sociological Inquiry*, 91(3): 625–646. https://doi.org/10.1111/soin.12390

Martin, C. (1996) *A Sourcebook on Feminist Theatre and Performance: On and Beyond the Stage*. London: Routledge.

Massumi, B. (2002) *Parables for the Virtual: Movement, Affect, Sensation.* Durham, NC: Duke University Press.

Mcqueen, P. (2011) *A Life Less Liminal? Issues of Inclusion and Recognition for Trans-Identities. Gender, Citizenship and Inclusion.* ECPR General Conference Reykjavík, 25–27 August 2011, https://ecpr.eu/Events/Event/PaperDetails/7075.

Meyer, J. H. and Land, R. (2005) Threshold Concepts and Troublesome Knowledge: Epistemological Considerations and a Conceptual Framework for Teaching and Learning. *Higher Education*, 49(3): 373–388. https://doi.org/10.1007/s10734-004-6779-5

Perger, N. (2022) Affective Obligations and Obliged Affections: Non-binary Youth and Affective (Re)orientations to Family. In: M. Kolehmainen, A. Lahti and K. Lahad (eds) *Affective Intimacies*. Manchester: Manchester University Press, pp 157–175.

Pile, S. (2010) Emotions and Affect in Recent Human Geography. *Transactions of the Institute of British Geographers*, 35(1): 5–20. https://doi.org/10.1111/j.1475-5661.2009.00368.x

Prosser, J. (1995) No Place like Home: The Transgendered Narrative of Leslie Feinberg's Stone Butch Blues. *MFS Modern Fiction Studies*, 41(3): 483–514. https://doi.org/10.1353/mfs.1995.0120

Rattray, J. (2016) Affective Dimensions of Liminality. *Educational Futures: Rethinking Theory and Practice*, 68: 67–76. https://doi.org/10.1007/978-94-6300-512-8_6

Robertson, N. (2018) The Power and Subjection of Liminality and Borderlands of Non-Binary Folx. *Gender Forum*, (69): 45–59.

Schmitz, S. and Ahmed, S. (2014) Affect/emotion: Orientation Matters. A Conversation between Sigrid Schmitz and Sara Ahmed. *Freiburger Zeitschrift Für GeschlechterStudien*, 22(2): 97–108. https://doi.org/10.3224/fzg.v20i2.17137

Sedgwick, E. K. (2006) *Touching Feeling: Affect, Pedagogy, Performativity.* Durham, NC: Duke University Press.

Segura, D. A. and Zavella, P. (2008) Gendered Borderlands. *Gender and Society*, 22(5): 537–544. https://doi.org/10.1177/0891243208321520

Tomkins, S. S. (1963) Introduction: Consciousness and Affect in Behaviorism and Psychoanalysis. In: *Affect, Imagery, Consciousness, Vol. 1: The Positive Affects*, 3–27. https://doi.org/10.1037/14351-001

Thrift, N. (2004) Intensities of Feeling: Towards a Spatial Politics of Affect. *Geografiska Annaler: Series B, Human Geography*, 86(1): 57–78. https://doi.org/10.1111/j.0435-3684.2004.00154.x

Weston, K. (1997) *Families We Choose: Lesbians, Gays, Kinship.* New York: Columbia University Press.

Yang, G. (2000) The Liminal Effects of Social Movements: Red Guards and the Transformation of Identity. *Sociological Forum,* 15(3): 379–406.

13

Unsettling Metronormativity: Locating Queer Youth in the Regions

Nicholas Hill, Katherine Johnson, Anna Hickey-Moody, Troy Innocent and Dan Harris

Introduction

Queer displacement is a central theme within the popular song 'Smalltown Boy', Bronski Beat's first single from their debut album. A young gay man stands alone on a platform waiting for a train to transport him to a more hopeful place; 'the wind and the rain on a sad and lonely face'. The feelings of pain, loss and hopelessness that permeate the song generate a powerful image of a generic small town inhospitable to queer life. The second verse and chorus ('run away, turn away') illustrate the concept of mobility within popular and academic understandings of the (im)possibility of queer life in regional and rural areas, 'the answers you seek will never be found at home, the love that you need will never be found at home'. To actualize a queer identity and find love, lesbian, gay, bisexual, trans, intersex, queer and asexual (LGBTIQA+) people are thought to move from the regions to safer and more accepting urban environments (Gorman-Murray, 2007; Cover et al, 2020). Reinforcing this narrative of displacement are cautionary tales of discrimination or violence against LGBTIQA+ who (mistakenly) try to settle or remain in the regions. From this metropolitan viewpoint, rural and regional areas are constructed as strange, distant and filled with populations hostile to LGBTIQA+ communities and dangerous individuals (Halberstam, 2005).

These distinct but complementary narratives – rural-to-urban journeys of queer self-discovery (Gorman-Murray, 2007) and the discrimination and violence that LGBTIQA+ are potentially subject to in regional areas – have contributed to what Halberstam (2005) terms a metronormative narrative. LGBTIQA+ people living outside of urban centres are constructed as 'sad' or 'lonely' or, alternatively, 'stuck in a place that they would leave if they only could', whereas queers who migrate to metropolitan centres are able to fully express their gender identity and sexuality (Halberstam, 2005: 36). Metronormativity significantly shapes LGBTIQA+ focused research, policy and service design, but falling outside this hegemonic frame are the complex ways that LGBTIQA+ people find a sense of place, forge connections and establish a sense of belonging in rural and regional areas. The invisibility of successful queer lives limits what affective possibilities are available to LGBTIQA+ youth in the regions when they try to locate themselves within their immediate surroundings and orient themselves toward the future (Esteban Munoz, 2009).

This chapter provides an overview of our attempt to work with LGBTIQA+ young people living in Kyneton, a small town in Victoria, Australia, to explore positive experiences of place, connection and belonging. Our initial idea was to co-create a prototype for a placed-based digital intervention that would disrupt pathologizing and risk-based framings of young queer lives in the regions and explore whether this could affectively transform perceptions of the regions as necessarily anti-queer by making available more positive and hopeful queer subject positions. Before providing an overview of the intervention and methodological approach, we examine scholarship that challenges the dominance of metronormativity and argue for the importance of understandings that promote affective connections. We conclude by highlighting the potential of digital place-based interventions to affectively transform perceptions of regional and rural areas as hostile and damaging for queer youth.

Locating queer youth in regions

Metronormative understandings of queer lives have conflated queer migration with rural-to-urban relocations and contributed to a problematic teleology whereby LGBTIQA+ individuals are viewed as emerging from the 'rural closet' within urban spaces (Gorman-Murray, 2007). This narrative trope has devalued the rural within the urban/rural binary (Halberstam, 2005: 38) and contributed to a metronormative bias within research (Conner and Okamura, 2022). Literature highlights how heteronormative culture impacts on LGBTIQA+ people located in the regions differently to queers living in urban areas and may contribute to rural queers downplaying their gender identity and/or sexuality (Armstrong et al, 2020; Conner and Okamura, 2022). Work focusing on LGBTIQA+ youth in regional and rural areas has contributed to a growing picture of queer youth in the regions as a vulnerable and marginalized group – at risk of mental health problems, educational drop-out, family exile and with limited access to both formal and informal support (compare Dahl et al, 2015; Bowman et al, 2020; Goldbach et al, 2023). While this deficit approach can be important for targeting policy development and service provision, the continuous reproduction of research highlighting the vulnerability of LGBTIQA+ youth in terms of discrimination, social exclusion and health and well-being disparities can become 'sticky' and restrictive in terms of subjectivity and transformation (see Zoli et al, 2024).

Pathologizing and deficit-based representations of rural queer youth limit the subject positions and affective conditions of liveability available to rurally located queer young people. Acknowledging this is important, because they are a dominant framework through which young LGBTIQA+ people become intelligible both to themselves and others (Butler, 2016). Within these critical or paranoid frames (Sedgwick, 2003), alternative outcomes for queer lives are rendered invisible, limiting the availability of more positive, hopeful and socially connected subject positions (Johnson, 2015). Narratives of isolation, marginalization and discrimination risk becoming a self-fulfilling prophecy because they largely constitute the affective and discursive resources available to LGBTIQA+ young people through which they locate themselves in the worlds they inhabit and consider their futures (Esteban Munoz, 2009). Geography continues to make a difference in terms of migration patterns (Thorsteinsson et al, 2022) and possibilities for support and affirmation of queer identities within small cities and regional towns, but more complex and nuanced understandings are required (Hulko and Horanes, 2018).

The need to diversify our understanding of the lives of rural queer youth is reinforced by a growing body of research questioning the cultural myth that young people growing up in regional and rural areas need to move way to actualize an LGBTIQA+ identity and find community (Cover et al, 2020; Marlin et al, 2022). Explorations of queer belonging and experiences of social inclusion/exclusion in rural and regional areas (Gorman-Murray et al, 2008; Brown et al, 2009) are generating insights into the diverse strategies that LGBTIQA+ people use to negotiate rural and regional settings, which include commuting to metropolitan centres to access queer communities, engaging in political activism and shifting self-conceptions toward non-sexualized aspects of self (Conner and Okamura, 2022: 1051). Regionally situated childhood experiences of younger LGBTIQA+ generations are no longer necessarily configured through concepts of closetedness, secrecy and isolation, and regional settings are 'not always exclusive, homophobic and violent, nor spaces in which families marginalise diversity' (Cover et al, 2020: 332). Rural and regional contexts can and do affect the development of LGBTIQA+ identities at this crucial developmental time in both positive and negative ways (Dahl et al, 2015; Johnson, 2015). Queer belonging in rural areas can be achieved through the development of behavioural competencies in identity management and adapting behaviour to suit local norms or via online communities, but negative perceptions of LGBTIQA+ social acceptance can inhibit belongingness and influence the way competencies, motivations and opportunities are enacted within local contexts (Marlin et al, 2022).

Examining the way LGBTIQA+ youth navigate regional and rural areas can produce new insights into the ways place, connection and belonging are enacted through everyday interactions and attachment to place. Anna Hickey-Moody (2016) suggests that the multiple 'little publics' to which young people belong should be taken seriously because it is through affective attachments, contestation and resistance that social value and citizenship is produced, enacted and sustained. This concept

> opens out and activates the politics that the term social inclusion signifies, but does so in a way that shifts the focus from including youth in a dominant paradigm to having the possibility of youth creating their own dialogic space which might speak back to a dominant paradigm, or might acquiesce. (Hickey-Moody, 2016: 68)

Embracing the idea that 'everyday theory qualitatively affects everyday knowledge and experience' (Sedgwick, 2003: 145) generates insights into the complex and ambivalent 'relations between the private and the public, the psychic and the social, and lived experience and social systems' (Cvetkovich, 2012: 25). As a practice, it transcends the limitations of examining the psychological and the social as separate spheres by turning toward the *psychosocial* – a transdisciplinary concept that understands inner psychological life and social conditions as mutually constitutive of subjectivity (Frosh, 2010: Johnson, 2015).

Vitalist concepts, within new materialist scholarship, also highlight the ongoing interaction between self and other, self and world and the interrelationship between bodies and things (Colebrook, 2010). Vitalism, according to Fuchs (2012), is the *feeling* of being alive, an intricate connection between the organic process of life, environment and subjective experience, all of which are critical to a sense of well-being. The notion of aliveness is similar but focuses on the feelings and processes of struggle that are central elements of LGBTIQA+ lived experience as they strive to achieve a liveable life in a heterosexist society with binary gender expectations (McGlynn et al, 2020). These concepts focus attention on what Kathleen Stewart terms 'ordinary affects':

> The ordinary is a shifting assemblage of practices and practical knowledge, a scene of liveness and exhaustion, a dream of escape or of the simple life. Ordinary affects are the varied, surging capacities

to affect and to be affected that give everyday life the quality of a continual motion of relations, scenes, contingencies, and emergencies. (Stewart, 2007: 1)

Exploring ordinary affects shifts attention away from rationality and toward sensation and contributes to a new form of politics centred on emotions and affect, what Katherine Johnson describes as affective activism (2015). In turn, affective activism can produce new resources that provide the basis for an 'alternative ethics of living' (Seigworth and Gregg, 2010: 12). It is possible to do this type of 'reparative work' (Cvetkovich, 2012) through enhancing the way we generate and interact with qualitative data (Moreno-Gabriel and Johnson, 2019) about the lives of young queer people living in regional areas. This requires attending to the associated affective and emotional registers involved in creating connection and belonging in place, and has the potential to generate new forms of activism and scholarship capable of transforming queer subjectivities and perceptions of regional areas as hostile to LGBTIQA+ youth.

Methodological contributions to a queer politics of belonging

In this section, we outline our research adventure and describe the way we worked with LGBTIQA+ youth to explore queer connections, attachments to place and experiences of belonging in Kyneton, a small regional town located approximately one hour from Melbourne, a large metropolitan city in Australia. The aim was to develop a concept/prototype for a digital place-based intervention that might achieve two outcomes: to unsettle the dominant representations of LGBTIQA+ youth located in regional and rural areas centred on vulnerability, discrimination and risk, capable of improving LGBTIQA+ youth well-being by generating new possibilities for place, connection and belonging. Katherine Johnson led a team of interdisciplinary researchers specializing in participatory research and the use of creative arts-based and narrative methods, including photo-voice, performance-based methods and digital animation, and the production of non-traditional research outputs such as digital apps, interactive games and art exhibitions. A 'digital intervention' was identified in advance as a possibility, based on the research team's expertise and increasing evidence suggesting LGBTIQA+ youth use online spaces to access and share knowledge (Byron and Hunt, 2017), create a sense of belonging (Cover et al, 2020; Marlin et al, 2022) and develop strategies for political action (Hanckel and Morris, 2014). Seven young people aged 18+ from a local LGBTIQA+ youth group were recruited in total, with participation ranging from two to seven participants across four workshops. Ethical approval for the project was granted by RMIT University's Human Research Ethics Committee (22291-07/19).

The research utilized the principles of co-design, an established method for working with (often marginalized) communities and social groups to transform institutions, communities and society, facilitate social inclusion and ensure outputs are meaningful and useful to end-users (Johnson and Martinez Guzman, 2013; Greenhalgh et al, 2016; Hill and Johnson, 2021). Following Anna Hickey-Moody and Marissa Willcox (2019: 5), the co-production of data and outputs can be thought of as 'the residue' of a successful qualitative research process that engaged LGBTIQA+ young people in 'collaborative making practices' through which they could 'think together about the future, about social values and community life'. LGBTIQA+ youth were positioned as co-researchers instead of being located as subjects of the research, and they actively contributed to shaping the project and output. We embraced the idea that participatory approaches are not singular recipes but, rather, principles guiding research with flexible methods (Ortiz Aragan and Castilo-Burguete, 2020). Together with our co-researchers, we experimented with various creative methods to co-design the place-based digital intervention.

Defining and developing the digital intervention

Following initial introductions and an activity exploring identity, the group settled on the idea of developing a concept for an interactive digital game that individuals could use via a phone-based application to explore Kyneton and access LGBTIQA+ stories of place, connection and belonging either in place or online (more information about the application is provided later). To construct and populate the map, creative methods were used to explore the situated and everyday knowledge held by our co-researchers. They facilitated new opportunities for exploring the experiences of place, connection and belonging as they emerged within and through the 'little publics' in which our co-researchers were located (Hickey-Moody, 2016; Fullagar et al, 2019) and generated new forms of data for exploring affective connections (Moreno-Gabriel and Johnson, 2019). It was through creative activities that our co-researchers' emotional investments and affective attachments to place could be explored and reorganized (Hickey-Moody, 2016: 8), and more 'e-motional or sensational or tactile' forms of agency produced (Cvetkovich, 2012: 21). Arts-based activities facilitated a more equitable relationship between the research team and participants through the co-creation of a shared output and provided diverse opportunities for engagement (Coeman and Hannes, 2017; Johnson and Reavey, 2022). The choice of methods was iterative, consultative and playful – guided equally by the research team's expertise and our co-researchers' interests and knowledge of the local area.

Across four workshops conducted in late 2019, a hybrid method involving a mapping exercise, photography, walking interviews and story-telling was developed to explore the local area and produce content for the application (Hill and Johnson, 2021). A local tourist map, walking tour and photographic exercise were used to generate a queer image of the town. Mapping as a research method created a dialogical space where we were able to collectively explore important landmarks, identify queer places and affectively and emotionally locate formative LGBTIQA+ experiences (McGrath et al, 2020). In addition, it provided a methodological space for exploring what a transformative intervention might look like in practice and how it would work. Once a queer topography of Kyneton was produced, story-telling exercises were used to document hidden queer narratives that had the potential of transforming perceptions of non-urban areas as necessarily homophobic, biphobic and transphobic (Bellamy, 2018: 688). In the following section, we describe the concept for the placed-based digital intervention and provide an example of how it would work.

Queer(y)ing place: 'Rainbow Rangers'

All except one of our co-researchers had grown up and attended a school in the local area (these are marked in Figure 13.1) and eagerly contributed ideas for the eventual design of the placed-based digital intervention. The concept evolved across the workshops through discussions about what digital platforms and applications the group used, how it might work in practice and user experience. The need for an informative resource that could be accessed safely and discreetly was balanced by our co-researchers' strong emphasis on ensuring that it could facilitate in-person connections. Online connections were considered to be helpful by the group, but one group member explained that without the possibility of meeting queer people in person, 'You're just surviving'. Group discussions focused on formative LGBTIQA+ experiences, where they had taken place and identifying queer-friendly spaces such as a local café, the bookstore and theatre. Marking these sites and hearing the narratives helped us to become more familiar with the town. Their attachments to place were complicated by where our co-researchers saw themselves living in the future. One of our co-researchers explained, "I kind of treat Kyneton as a home base to branch out to other places. It's not a place that I would want to settle in for the rest of my future."

Experiences of place, connection and belonging were shaped by the everyday realities of small-town life, especially the lack of anonymity. Many group members were 'out' to only a few trusted friends and family members, and the limited number of physical spaces where they felt safe and comfortable to be themselves were noted. The risk of being seen was noted

Figure 13.1: Conceptual rendering of 'Rainbow Rangers'

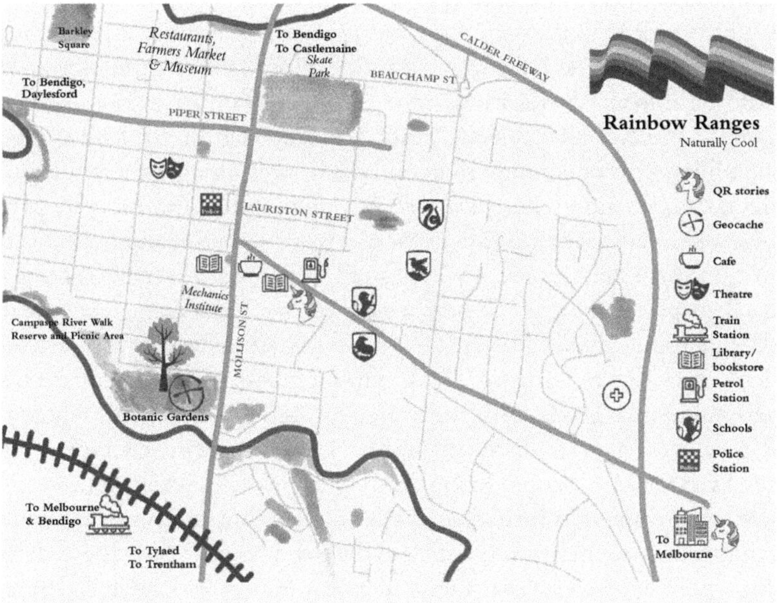

Source: Nicholas Hill, Katherine Johnson, Anna Hickey-Moody, Troy Innocent, Dan Harris

by one co-researcher: "Everywhere you go there is a chance you will be seen. And even if you're not in an obvious [queer] meeting group, you're still with people that may act different and that is enough." Despite this wariness, our co-researchers continued to seek queer connections within the local area.

Some of group had moved to Melbourne after completing secondary school for further study and/or work, yet Kyneton was still considered a home base they remained attached to. Each group member was engaged in either some form of work, study or a voluntary role, many were involved in theatre and music and many were involved in community activism. Our co-researchers enthusiastically described their involvement in LGBTIQA+ events and community-based activities, such as LGBTIQA+ camps, queer formals, flag-raising ceremonies and queer activism. The highs and lows of engagement in local protests and activism related to Australia's 2017 Marriage Equality Plebiscite, which caused significant distress within LGBTIQA+ communities (Ecker et al, 2019), were prominent throughout discussions. Isolation from other LGBTIQA+ young people was a key concern expressed across the four workshops and shaped the final concept design. One of our youth co-researchers put it this way: "Being queer and rural ... [is] less talked about, there's less people in general, so it's less likely that there's going to be other queer people in your community and town." This social isolation was more pronounced for one group member who was also geographically isolated and reliant on their parents, siblings and irregular public buses to access what limited queer life existed in the town.

Within workshops two and three, our co-researchers were invited through various arts-based methods to creatively explore how we might transform hostile or unfriendly locations into more LGBTIQA+ friendly spaces. A mapping exercise helped the group to locate significant landmarks and situate formative queer experiences. During a walking tour, photographs were taken of sites where important queer public events and private experiences had occurred. These locations were then marked on a tourist map. Sites included the street where a rally in support of marriage equality had taken place, the place where a non-binary participant received their first binder and the local theatre – which was viewed as very 'queer' and a welcoming space for many of the group members. They also noted places where some had 'come out', and were affirmed and accepted, where they had gone on dates or shared their first kiss, and joyful encounters with older LGBTIQA+ people living in the area. Serendipitous and everyday encounters were transformative, with a queer life in Kyneton suddenly becoming possible in that moment and the town less isolating. It was decided to incorporate these everyday but hidden stories into a conceptual design for a map-based digital game, an interactive application that bridged the divide between the virtual and the physical and provided opportunities for players to meet other users – in person or online – while simultaneously offering a sense of safety and security.

Drawing on the playfulness associated with queer culture, the game was to be called Rainbow Rangers. The eventual design centred on an interactive map accessed through Instagram and incorporated Discord, an instant messaging social platform, and geocaching activities. Figure 13.1 is a conceptual rendering of the application, which aimed to promote queer inclusion and affectively transform perceptions of regional areas as antithetical to queer life. The final design was based on a local tourist map of Kyneton and the surrounding area with significant queer landmarks such as the local theatre, LGBTIQA+ owned cafes and bookstore and local schools marked. The use of alternative symbols, bright colours and the insertion of the rainbow transformed the former gold-mining town into a welcoming and inclusive space. The symbols used to locate the schools our co-researchers had variously attended were inspired by the different houses of Hogwarts, the school where the Harry Potter book series takes place. Echoing the tracks featured in the video from 'Smalltown Boy', the train station remains a vital queer site. Our co-researchers said they often visited Melbourne to connect with other LGBTIQA+ young people and go on dates. One co-researcher remarked that dating options were limited in Kyneton: "Unless you find someone through word of mouth, you just don't". When asked where they would take someone on a first date, the group replied in unison,

"Melbourne". The intimacy, visibility and vulnerability of going on a date in one of the local venues was felt to be too personal: "It's welcoming them in". This adds to research challenging metronormativity by illustrating the permeability of the urban/rural divide (McGlynn et al, 2020; Conner and Okamura, 2022) and the importance of the trainline as a form of mobility both out of and into Kyneton.

The place-based game was designed so that those who wanted to explore Kyneton in person or with friends could follow the interactive map and locate the geocache site. A QR code would then be found within that site. Scanning it would unlock an audio recording of a queer story that had occurred within that place. Nondescript carparks, the local gardens, buildings, would be suddenly transformed through listening to hidden and surprising queer stories of place, connection and belonging. Individuals would be able to affectively and emotionally connect with queer lives in Kyneton and exchange stories with other users. Revealing these hidden LGBTIQA+ stories made available different possibilities and alternative visions of a future in a place that might be perceived as scary, unwelcoming or hostile to young queer people. Interactive map-based digital applications, such as this, work by shifting meaning and representation through a 'blending of the world-as-it-is to the world-as-it-could-be' and, in doing so, reveal often hidden meanings or invisible experiences and, by extension, renew and remake place through interaction (Innocent and Leorke, 2019: 28–29). In hearing queer stories of place, connection and belonging, it becomes possible to generate hopeful and positive representations of LGBTIQA+ young people living in regional areas and transform perceptions of the regions as necessarily hostile to queer youth.

Stories of hopeful moments and positive encounters were shared by our co-researchers with the intention that these might populate the digital intervention in the future. An example is pictured in Figure 13.2, a site where several of our group members found acceptance and established queer connections. Individuals participating in the place-based game could visit the site and, using geocache, listen to a young person's discovery of a welcoming space and their journey toward self-acceptance.

This foreboding grey building is located on a long street near the centre of Kyneton. The heavy stone, stained-glass windows and wrought-iron fence allude to its former life as a Congregational church. It has since been repurposed as a much-cherished theatre and is rumoured to be haunted by a ghost. The building's unwelcoming appearance does not hint at the warmth, acceptance and affirmation one of our co-researchers found inside.

'Kyneton Theatre Company has a special place in my heart. I started doing theatre back in 2015 and KTC was the first community theatre I was a part of. An ex-partner of mine invited me one night to hang with them while they rehearsed. At the time I was questioning my sexual identity and was struggling to come to terms with who I was as I was brought up through media portraying LGBTQI people as outcasts. But after that rehearsal I found like, I found a place that was openly accepting, and a place where I was fully accepted without question. I joined that show as a band member. And since then, I have continued to pursue theatre and have moved to being on stage, both through school shows and community theatre. A few years later I was comfortable enough to come out to a few friends. And then I was in a show where some of the members were open and I decided to come out to my second family as asexual.'

This story begins by highlighting the significance of the theatre company for affective transformation. Our co-researcher explains how he had initially

Figure 13.2: Kyneton's Bluestone Theatre

Source: Nicholas Hill, Katherine Johnson, Anna Hickey-Moody, Troy Innocent, Dan Harris

struggled with his sexuality, and links this to harmful media portrayals of LGBTIQA+ people which marginalize queer youth and position them in pathological ways. In contrast to "LGBTQI" being "outcasts", the theatre was a space where he felt "fully accepted" for the first time and able to be comfortable and safely explore his sexuality. Through becoming more immersed in the theatre and forming new social connections, he slowly felt comfortable enough to come out to a few friends. Eventually, through a show where some members were open about their LGBTIQA+ identity, he said he gained the self-confidence to "come out to my second family as asexual". Our co-researcher's story resists and challenges hegemonic narratives that position young queer people in the regions as necessarily isolated and suffering poor mental health. It demonstrates how LGBTIQA+ young people can and do find a home and places in the regions where they are able to form social connections and live affirmatively.

Conclusion

Our methodological adventure in working with LGBTIQA+ young people living in a small regional town to co-create a place-based intervention unsettles metronormative assumptions within research, policy and service provision. Through working alongside queer youth and using creative methods to explore, resist and transform the 'little publics' (Hickey-Moody, 2016) to which they belonged, new representations of LGBTIQA+ youth in the regions and queer ways of seeing and interacting with a small town were generated. Our co-researchers' stories and contributions highlight the importance of home and place for queer youth and the complex ways they navigate small-town life and think about the future. The allure of the big city for LGBTIQA+ youth growing up in regional areas remains, but, unlike the sad, lonely figure within the song 'Smalltown Boy' that opened this chapter, strong queer attachments to place are evident. Feelings of isolation, fear and vulnerability were evident within the contributions offered by our co-researchers. But alongside these were hopeful, intimate and joyful descriptions of place. The complex and ambivalent emotions evident within our co-researchers' stories of growing up in Kyneton highlight the value of attending to the 'everyday knowledge and experience' (Sedgwick, 2003) that LGBTIQA+ youth accumulate as they struggle to make a life for themselves in a place that has traditionally been understood as holding few possibilities for queer life.

Our creative and playful collaboration with regionally located LGBTIQA+ young people highlights new possibilities for developing interventions aimed at supporting the mental health and well-being of queer youth in the regions. The place-based game queers Kyneton by allowing users to explore and engage with hidden queer stories of serendipitous encounters and joyful connections both online and in the place where they occurred. New subject positions are made available through the application that LGBTIQA+ young people can draw on when attempting to locate themselves in the world and think about their future. It has the potential to reduce the isolation that young queer people within the regions may feel by allowing users to affectively connect to the stories of other users and, in addition, it provides a space where they can connect digitally and, importantly for our co-researchers, in person. Through the place-based game, metronormative assumptions about rural life are affectively transformed and the regions become a place of hope and possibility for young LGBTIQA+ people. Small-town life may involve struggle, but understanding and drawing on the ways in which LGBTIQA+ young people find ways to belong and form connections generates new affective resources for queer(y)ing place.

References

Armstrong, E., Coleman, T., Lewis, N. M., Coulombe, S., Wilson, C. L., Woodford, M. R., ... and Travers, R. (2020) Travelling for Sex, Attending Gay-specific Venues, and HIV-related Sexual Risk among Men Who Have Sex with Men in Ontario, Canada. *The Canadian Journal of Human Sexuality*, 29(3): 380–391.

Bellamy, R. (2018) Creative Health Promotion Methods for Young LGBTIQA+ People. *Health Education Journal*, 77(6): 680–691. https://doi.org/10.1177/0017896917753454.

Bowman, S., Easpaig, B. and Fox, R. (2020) Virtually Caring: A Qualitative Study of Internet-based Mental Health Services for LGBT Young Adults in Rural Australia. *Rural and Remote Health*, 20(1). DOI: 10.22605/RRH5448.

Browne, K. Lim, J. and Brown, G. (eds) (2009) *Geographies of Sexualities: Theory, Practices and Politics*. Farnham: Ashgate.

Butler, J. (2016) *Frames of War: When is Life Grievable?* Brooklyn, NY: Verso.

Byron, P. and Hunt, J. (2017) 'That Happened to Me Too': Young People's Informal Knowledge of Diverse Gender and Sexualities. *Sex Education: Sexuality, Society and Learning*, 17(3): 319–332.

Coemans, S. and Hannes, K. (2017) Researchers under the Spell of the Arts: Two Decades of Using Arts-based Methods in Community-based Research with vulnerable populations. *Educational Research Review*, 22: 34–49.

Colebrook, C. (2010) Queer Vitalism. *New Formations*, 68(68): 77–92.

Conner, C. and Okamura, D. (2022) Queer Expectations: An Empirical Critique of Rural LGBT+ Narratives. *Sexualities*, 25(8): 1040–1057.

Cover, R., Aggleton, P., Rasmussen, M. and Marshall, D. (2020) The Myth of LGBTQ Mobilities: Framing the Lives of Gender- and Sexually Diverse Australians between Regional and Urban Contexts. *Culture, Health and Sexuality*, 22(3): 321–335.

Cvetkovich, A. (2012) *Depression: A Public Feeling*. Durham, NC: Duke University Press.

Dahl, A. L., Scott, R. K. and Peace, Z. (2015) Trials and Triumph: Lesbian and Gay Young Adults Raised in a Rural Context. *Social Sciences*, 4(4): 925–939.

Ecker, S., Riggle, E., Rostosky, S. and Byrnes, J. (2019) Impact of the Australian Marriage Equality Postal Survey and Debate on Psychological Distress among Lesbian, Gay, Bisexual, Transgender, Intersex and Queer/ Questioning People and Allies. *Australian Journal of Psychology*, 71(3): 285– 295. DOI: 10.1111/ajpy.12245.

Esteban Muñoz, J. (2009) *Cruising Utopia: The Then and There of Queer Futurity*. London: New York University Press.

Frosh, S. (2010) *Psychoanalysis Outside the Clinic: Interventions in Psychosocial Studies*. New York: Bloomsbury Publishing.

Fuchs, T. (2012) The feeling of being alive: Organic foundations of self-awareness. In: J. Fingerhut and S. Marienberg (eds) *Feelings of Being Alive*. New York: De Gruyter, pp 149–165.

Fullagar, S., O'Brien, W. and Pavlidis, A. (2019) *Feminism and a Vital Politics of Depression and Recovery*. London: Palgrave Macmillan.

Goldbach, J., Parra, L., O'Brien, R., Rhoades, H. and Schrager, S. (2023) Explaining Behavioral Health Differences in Urban and Rural Sexual Minority Adolescents. *Journal of Rural Health*. 39: 262–271. DOI: 10.1111/ jrh.12706.

Gorman-Murray, A. (2007) Rethinking Queer Migration through the Body. *Social and Cultural Geography*. 8(1): 105–121.

Gorman-Murray, A., Waitt, G. and Gibson, C. (2008) A Queer Country? A Case Study of the Politics of Gay/Lesbian Belonging in an Australian Country Town. *Australian Geographer*, 39(2): 171–191.

Greenhalgh, T., Jackson, C., Shaw, S. and Janamian, T. (2016) Achieving Research Impact through Co-creation in Community-Based Health Services: Literature Review and Case Study. *The Millbank Quarterly*, 94(2): 392–429.

Halberstam, J. J. (2005) *In a Queer Time and Place: Transgender Bodies, Subcultural Lives*. New York: New York University Press.

Hanckel, B. and Morris, A. (2014) Finding Community and Contesting Heteronormativity: Queer Young People's Engagement in an Australian Online Community, *Journal of Youth Studies*, 17(7): 872–886, DOI: 10.1080/ 13676261.2013.878792.

Hickey-Moody, A. (2016) Youth Agency and Adult Influence: A Critical Revision of Little Publics. *Review of Education, Pedagogy, and Cultural Studies*, 38(1): 58–72.

Hickey-Moody, A. and Willcox, M. (2019) Entanglements of Difference as Community Togetherness: Faith, Art and Feminism. *Social Sciences*, 8: np.

Hill, N. and Johnson, K. (2021) Inclusive Research with LGBTIQA+ Groups: Reflections on the Participatory Turn. In: P. Liamputtong (ed) *Handbook of Social Inclusion: Research and Practices in Health and Social Sciences*. Cham: Springer International Publishing, pp 1–21.

Hulko, W. and Hovanes, J. (2018) Intersectionality in the Lives of LGBTQ Youth: Identifying as LGBTQ and Finding Community in Small Cities and Rural Towns. *Journal of Homosexuality*, 65(4): 427–455.

Innocent, T. and Leorke, D. (2019) Heightened Intensity: Reflecting on Player Experiences in Wayfinder Live. *Convergence*, 25(1): 18–39

Johnson, K. (2015) *Sexuality: A Psychosocial Manifesto*. Cambridge: Polity.

Johnson, K. and Martínez Guzmán, A. (2013) Rethinking Concepts in Participatory Action Research and Their Potential for Social Transformation: Post-structuralist Informed Methodological Reflections from LGBT and Trans-Collective Projects. *Journal of Community and Applied Social Psychology*, 23: 405–419.

Johnson, K. and Reavey, P. (2022) Designing Qualitative Research for Working with Visual Data: Insights from Psychology. In: *The SAGE Handbook of Qualitative Research Design*. London: SAGE.

Marlin, L., Lewis, C. and McLaren, S. (2022) *'Being Able to Be Yourself'*: A Qualitative Exploration of How Queer Emerging Adults Experience a Sense of Belonging in Rural Australia, *Journal of Homosexuality*. DOI: 10.1080/00918369.2022.2092806.

McGlynn, N., Browne, K., Banerjea, N., Biswas, R., Banerjee, R. S. and Bakshi, L. (2020) More than Happiness: Aliveness and Struggle in Lesbian, Gay, Bisexual, Trans and Queer Lives. *Sexualities*, 23(7): 1113–1134.

McGrath, L., Mullarkey, S. and Reavey, P. (2020) Building Visual Worlds: Using Maps in Qualitative Psychological Research on Affect and Emotion. *Qualitative Research in Psychology*, 17(1): 75–97.

Moreno-Gabriel, E. and Johnson, K. (2019) Affect and the Reparative Turn: Repairing Qualitative Analysis. *Qualitative Research in Psychology*, 17(1): 98–120.

Ortiz Aragan, A. and Castilo-Burguete, M. (2020) Introduction to Practices. In H. Bradbury (ed) *The Sage Handbook of Action Research*. London: SAGE, pp 13–16.

Sedgwick, E. (2003) *Touching Feeling: Affect, Pedagogy, Performativity*. Durham, NC: Duke University Press.

Seigworth, G. and Gregg, M. (2010) An Inventory of Shimmers. In: M. Gregg and G. J. Seigworth (eds) *The Affect Theory Reader*. Durham, NC: Duke University Press, pp 138–157.

Stewart, K. (2007) *Ordinary Affects*. Durham, NC: Duke University Press.

Thorsteinsson, E. B., Bjarnason, T., Loi, N. M. and Arnarsson, A. M. (2022) Sexual Orientation and Migration Intentions among Rural, Exurban and Urban Adolescents in Iceland. *Culture, Health and Sexuality*, 24(1): 31–47.

Zoli, A., Johnson, K. and Hazenberg, E. (forthcoming) Unsettling Vulnerability: Queer and Feminist Interventions. *Feminism and Psychology*.

14

Landscape, Gender and Belonging: Male Manual Workers in a UK Seaside Town

Ruth Simpson and Rachel Morgan

Introduction

This chapter explores the 'gendering' of landscape and how it is claimed through affective experiences of belonging. It draws on a recent study of male manual workers in a 'struggling' UK seaside town: Hastings in East Sussex. It highlights how the physical and cultural attributes of landscape help to generate attitudes of inclusion and exclusion through, in part, the claiming of landscape as well as through ideologically charged and dominant ways of seeing and defining. From this perspective, landscape is not just as a set of physical and topographical features but is also gendered, marked by history and culturally produced, bound up with personal, interpersonal and affective experience. As a mid-sized coastal town on the south coast of the UK, Hastings is both a place of 'leisure and pleasure', through a reliance on its attraction as a tourist destination, and a working town. In terms of the latter, an occupational heritage oriented around fishing has helped to create a 'psychic economy' (Nayak, 2003) based on gendered notions of manual work.

We accordingly focus on gendered understandings of landscape and highlight how landscape generates affective belonging, exclusive of 'outsiders' and which relates to how it is both inhabited and claimed. This is to see affect as the active outcome of an interaction or encounter (Deleuze, 1998; Thrift, 2004) and to examine how immersion in landscape can produce gendered, classed and racialized 'affective dispositions'—where these dispositions capture thoughts, values, perceptions, moods and 'states of being' (Hemmings, 2005). Rather than a single, identifiable emotion, affects can therefore be seen as 'broad tendencies' (Thrift, 2004) or lines of force that generate embodied practice through the dynamic processes of interaction. Inherently unstable and uncertain, affect resides in bodies as we encounter others, while also flowing between bodies as a transpersonal capacity (Anderson, 2006). This circulation of affect both shapes and is shaped by social relations. Here, context is a vital element in the constitution of affect as people engage with their environment (Sedgwick, 2003). This involves deeply felt resonances that exist 'in-between' individuals through their encounters with each other (Gregg and Seigworth, 2010) and in-between individuals and place. With an orientation to landscape as an assemblage of physical features and as interrelational and culturally produced, and with an emphasis on affect's bodily capacity, relationality and transmissibility, we consider how encounters with landscape can engender a sense of belonging that is drawn on gendered as well as classed and racialized lines.

Our choice of location reflects the marginal and deprived status of many seaside towns in the UK, situated on the geographical as well as social and economic periphery and where there is a widening economic gap between coastal and non-coastal communities (Corfe, 2017). These disadvantages can be keenly felt by white, working-class, year-round residents of seaside towns (McDowell and Bonner-Thompson, 2020) in a context where members of this group face particular challenges, given the few opportunities for secure work (Burdsey, 2011; Wenham, 2020). This situation is exacerbated for less-skilled men, the focus of our study, who are often seen as less eligible than women in terms of aptitude for the customer-facing service employment that can dominate these towns.

Hastings has relied on its attraction as a tourist destination, where a heritage based on a famous battle in 1066 (precipitating the Norman conquest of England), a historic castle and an attractive beach have proved popular with day-trippers and international tourists. The town is home to one of the oldest fishing ports in the UK and a small, beach-launched fishing industry, located in the Old Town (which was the extent of the town prior to the 19th century), still operates from a specific part of the seafront named The Stade. Pride in its historic and cultural legacy, in a lively summer trade and a meaningfulness based on an ancient invasion, tradition and the sea partly define the town. Nevertheless, Hastings is among the ten most deprived towns in England (Corfe, 2017), characterized by low relative income levels (Hastings.gov.uk, 2019), high unemployment (a higher-than-average proportion of jobs are in manual and 'low level' service occupations such as construction, hotel, catering and retail, much of which is seasonal and insecure) and with over a fifth of the population identified as living in deprived households (Department for Communities and Local Government, 2015).

The promenade is fronted by large, five-storey Victorian houses, many of which comprise houses of multiple occupation (HMOs) (Hastings was designated as an asylum seekers dispersal area in 2000). In good weather, tourists throng the Old Town, with its small shops and cafes, visit the fishing museum near The Stade or ride the two funicular railways to the East and West cliffs – more affluent areas that have panoramic views across the sea. While there is evidence of 'gentrification' as smaller Victorian terraced houses are restored by newly arrived middle classes, there is also visible social deprivation, with rough sleepers occupying the dingy underpass that connects the main square with the seafront, and an endemic drink and drug problem that is played out at night in specific areas of the town.

The chapter draws in part on data from an ethnographic study, conducted in 2018–19, of the work experiences of 25 male manual workers in Hastings (see Simpson et al, 2022). Participants were employed or had recently been employed in various forms of manual work, including building and construction (for example, as roofer, scaffolder or 'general labourer'), decorating, gardening and warehouse work. Over half of the participants had been born in or around Hastings, with the remainder having moved to the town as a child or in early adulthood. All were white – a deliberate sampling strategy to reflect not only the demographics of the area (95 per cent white) but also the tendency for white working-class to be a particularly disadvantaged group (Skeggs, 2004; McDowell et al, 2020).

Belonging, affect and gender

As a key dynamic within the affective-geographical experience, belonging relates to our physical involvement with our environment and our embodied 'being in the world' through an emotional attachment to and identification with a bounded space (Emery, 2018). As Savage et al (2005) point out, against contemporary notions of fluidity, mobility and the reification of 'uprootedness', local attachments are still crucial as people seek a way of 'being at home' in an unstable world. Place and locality accordingly remain implicated in the formation of belonging in the face of transformations and instabilities through, for example, cosmopolitanism, mass movements and changing territorial identities.

As Emery (2018) points out, belonging can be seen as an 'assemblage' of emotions producing an affective state based on mutually constitutive notions of attachment, loyalty, solidarity and sense of affinity. With a focus on attachment, Probyn (1996) sees belonging as grounded in a strong desire (a yearning) for connection to both people and places – processes that privilege, in affective terms, intimacy and proximity. As such, it is a particularly powerful 'mode of subjectification', designating a profoundly affective 'manner of being' in terms of how people make sense of their lives. It is both individual and collective, emanating from the affective processes whereby a person becomes included in a 'placed' collective and identifies with it. However, for Probyn, belonging is an ongoing and essentially incomplete quest. It is 'out of reach' in that one can never truly feel, with certainty and security, that one belongs. Rather than being an 'end state' or finality, this is to see belonging as 'a tenacious and fragile desire that is ... increasingly performed in the knowledge of the impossibility of ever really and truly belonging' (Probyn, 1996: 8). Belonging is therefore an ongoing, uncertain process and may be evoked through personal and collective memories and emotional attachment to collectively shared place histories and experiences (Emery, 2018) as well as through boundary making based in part on gendered, classed, raced manifestations of inclusion and exclusion.

The latter suggests that belonging is partly formed in an intersectional context along multiple axes of difference. Here, the white working class has been associated with a spatial 'fixidity' in experiences of belonging that relates strongly to family ties (Simpson et al, 2022). By contrast, Savage et al (2005) refer to 'elective belonging', whereby the middle class select places to live before putting down roots in order to 'be at home' in a fluid world. As they argue, places become 'sites for performing identities', where the mostly mobile middle class 'attach their own biographies to their "chosen" residential location' (Savage et al, 2005: 29). Gender, too, can mean a different relationship with place (Savage et al, 2005) through social relations of inclusion and exclusion (McDowell, 1999). Here women are often seen as being less mobile than men and more rooted in family and locality (Skeggs, 2004), while working-class men, as Nayak (2003) found, are more likely to focus on waged labour as a source of belonging. This labour can form part of the 'psychic economy' of a region, creating a sense of belonging that is manifest in folklore, kinship histories, pasttimes and tradition. Class, race and gender therefore have a crucial role to play in understanding the complex connections between place, affect and belonging.

Landscape and affect

In its conventional usage, landscape refers to the shape and topography of a piece of land that is viewed from a particular point, thus incorporating the physical and material with a notion of the visual in terms of how landscape is seen (Cresswell, 2015). This places emphasis on the solidity of landscape as well as the role of the visual gaze. However, rather than seeing the figure gazing on the landscape as an a priori self, projecting a structure of meanings, recent accounts place emphasis on the phenomenological processes whereby landscape is bound up with affective experience through which the self is 'assembled and performed' (Cresswell, 2015). Thus, as a 'set of relational places', landscape has been defined as '"the world out there" as understood, experienced, and engaged with through human consciousness and active involvement' (Bender, 2006: 303). This emphasizes landscape both as an assemblage of physical and structural features and as affective, interrelational and culturally produced.

Such understandings highlight the significance of affective experiences and forces that exist 'inbetween' (Massumi, 2002; Gregg and Seigworth, 2010) as we move through, experience and connect with our environment. A 'geography of the inbetween' (Pile, 2010), in our terms the encounter between people and between people and landscape, can be seen as an intersubjective affective flow which, through its intensity, generates a particular mode of being and sets of dispositions. Landscape, therefore, is not just part of a symbolically charged, aestheticized world of views and vistas but also, through its consumption, a locus of identity formation (Bender, 2002) and a dynamic collective activity that, marked by history, is 'enmeshed' within the processes that shape how the world is organized and understood. As both an assemblage of physical features and a 'network of cultural codes' (Darby, 2000), landscape is differentially experienced and subjectively 'in the making'. This is a generative process of bodily involvement and immersion where selfhood is reciprocally solicited through 'landscape sensibilities' (Wylie, 2009) – processes involving perspectives, practices, imaginations, memories, textures and feelings. This is to place emphasis on bodily presence and the central role of landscape in constructions of self, identity, community and belonging.

As Darby (2000) notes, the coordinates of geography, class and gender intersect through a network of inclusions and exclusions that relate in part to how landscape is claimed, as well as through the conventional orientation to vistas, to ideologically charged and dominant ways of seeing and defining. In his book *The Wild Places*, the writer and academic Robert MacFarlane (2007), known for his books on landscape, nature and place, sought through his travels around remote areas of the UK to experience and construct an understanding of 'wilding'. As Jamie (2008) notes in her book review, this brought out a 'mix of class, gender and ethnic tension' as she imagined MacFarlane, as a middle-class white male, jumping 'on a sleeper train … with the declared intention of seeking wild places'. In this way, the wild and the remote are, in her words, claimed 'by the educated middle classes on a spiritual quest'. Drawing on narratives of masculine heroism and based on intrepid and extreme experiences involving sleeping outdoors, swimming in freezing waters, climbing and walking by day and night, MacFarlane arguably constructs (and claims) 'the wild' as a white, middle-class, masculine domain.

This resonates with Zukin's (1991) conceptualization of *authenticity* as a source of claim and belonging, suggesting a 'right' and affiliation to a landscape that is exclusive of outsiders. This can, in some contexts, be cultivated through long-term residency, local knowledge, use and habit where the materiality of landscape is imbued with meaning, acting as anchors in a changing environment. As Zukin argues, landscape is accordingly

Figure 14.1: Hastings Pier

Source: Ruth Simpson

conceptualized as an 'architecture' of social relations based on gender, class and race, shaped by dynamics of power as seen, for example, in struggles over the control of space where priority is often given to dominant groups. Thus, the new Hastings Pier, which reopened in 2016 after being damaged by fire, has been the source of dispute between longer-term (mainly working-class) residents who prioritized traditional pier attractions (fairground, gaming machines, fishing rights) and the new owner and council leaders, supported by an emerging middle class, who sought a performance and cultural venue (Moore, 2019). The resultant modernist and sparse structure (Figure 14.1) received scathing reaction from long-term residents, being described by one as a 'landing strip'.

This points to landscape as a 'field of impact' and as an 'ensemble' of material, social and affective practices (Zukin, 1991) where, as Kraak and Kenway (2002) point out, dominant groupings imprint their values on landscape and space, with profound implications for experiences of belonging.

Constructions of belonging in Hastings

For Massey (2006), landscape and its histories can be evoked in creating a cultural identity and a sense of belonging. Landscape accordingly 'talks back' (Bender, 2006) through subjective experience, engaging our senses and emotions. We highlight here the significance of Hastings' male, occupational history, particularly in relation to meanings created around harbour towns, to the sensory pleasures of living on the coast and to a sense of belonging where long-term residents claim cultural ownership of place, to the exclusion of outsiders.

Occupational legacy and belonging

'You've got four or five scaffold companies, you've got roofers, you've got decorators, you've got electricians, plumbers and oh, you need something done, ask Brian, phone Bobby, ask Phil. Yeah, that's where your community comes in' (Ray, roofer)

Described by one participant as a 'working town', value was placed on Hastings' characteristic reliance on manual labour where geographical separateness and poor transport connectivity were implicated in a 'place dependence' oriented around the need for local work. 'Pockets' of factory work outside the town, a substantial building and decorating trade (assisted by conversions, middle-class renovations and ongoing salt damage from the sea) and some tourist-related seasonal work associated with a coastal town are principal sources of local male employment, and finding local work within the area was a source of struggle and a preoccupation for all interviewees. Faith and pride in manual work supported a strong working-class community where knowledge of available work, as Ray intimates in the quote, was shared through 'unofficial' networks including word of mouth and 'pub talk' for odd jobs in gardening, decorating or the building trade.

An occupational heritage oriented around working with the sea, where a still thriving fishing industry offered an identity and distinctiveness, also contributed to a collective ethic of hard graft ('proper work' in Nick's terms). "Yeah. I think it's lovely down (at The Stade), the boats with the birds. In real life it's (fishing) ... real hard and I think people in power and in governments are really out of touch. If you've never done work, proper work ..." (Nick, painter/decorator). "I always liked the sea and I wanted to go fishing, before I left school, I used to go down to the fishing dock in Hastings and help the fishing boats, it's a very close-knit family thing" (Neil, plasterer). Here, the fishing environment (the presence of boats and fishing gear on The Stade and the iconic 'net shops' used for storage) and an occupational heritage that has passed largely down the male line creates a distinctive place character as well as shaping individual and collective identities. From Neil's account earlier, fishing is a "close-knit family thing", with families in the Old Town, the location of much of the industry, reportedly going back generations in terms of their involvement in the trade.

For Neil, fishing carries a particular localized authenticity – "it's real life ... it's real hard" – based on notions of masculine endurance, reflecting its status as a difficult and dangerous way of life. Fishing therefore is constructed as a gendered activity enabling displays of masculine prowess that forms part of the cultural inheritance of the area. As one participant commented of the local fishing industry, "it's in our genes", signifying how shared dispositions (for example, pride in physical skill and endurance; knowledge of the sea; satisfaction from a strong, male working-class community) are internalized within the body and, in relational terms, are passed on from one generation to the next.

In fact, none of the participants was involved in the fishing industry but it occupied a privileged position in the region and its cultural values were internalized, nevertheless. As Nick and Neil reveal earlier, watching the daily activity of fishing as boats come in and out of The Stade is a source of satisfaction, indicative of a shared cultural practice. The physical setting of The Stade and surrounding beach, as well as the physical environment (sea, weather, tides, coastal landscape), are therefore important in the construction of individual and collective identities. Many participants identified with the fishing community in terms of the challenges faced by the industry, displaying a detailed knowledge of regulations and quotas and the impact on the trade. This suggests that the meanings attached to landscape and its centrality as an arena for social relations and collective belonging contribute to the circulation of a particular classed and gendered set of 'affective dispositions'. As Acott and Urquhart (2014) found in their study of the industry, these dispositions form part of the 'psychic economy' (Nayak, 2003) of a region and can strongly appeal even to those who are not directly involved. The privileging of fishing as a source of prized, male manual work can accordingly be internalized through a value placed on 'hard graft' and displaced into other cultural activities, including recreational fishing (discussed later) and observing the boats coming in and out of The Stade (Figure 14.2).

Sensory experiences of landscape

In terms of the foregoing, all participants spent valued leisure time in coastal activities: fishing, walking, swimming or simply spending time on the beach (as Ron commented, "I like coming down here and watching the sea come in and out on the cliffs"). Fishing was a particularly important recreational activity and the local angling club (located close to The Stade) was a key part of the community, enabling and supporting the practice of a largely masculine skill. Recreational fishing enabled men to perform masculinity in a non-work setting and to engage in an activity that, in the context of work, holds a privileged cultural position in the area. "I mean I fish, I don't go out

Figure 14.2: Fishing boats on The Stade

Source: Ruth Simpson

in boats yeah, but a lot of people because they are by the sea, they fish and you meet the same faces when you go down to the beach" (Andy, bricklayer).

Pubs in the Old Town offered an additional forum for social cohesion, an important generator of social and cultural values around physical work and community. Landscape therefore can be seen to drive a series of territorial relations that, drawing on Acott and Urquhart's study of fishing communities, are 'woven into the material and phenomenological worlds' (Acott and Urquhart, 2014: 260).

An appreciation of the beauty of Hastings and of the surrounding countryside ran through all accounts, seen to have restorative benefits (in the words of one participant: "I like the therapeutic aspect") and compensating in part for the difficulties of finding secure work. "There's not much work, but there's the sea …!" "It's beautiful isn't it … and the hills around. I love it." "You can smell it (the sea), yeah."

Landscape accordingly offers active, physical interactions through leisure activities, and sensory, embodied, aesthetic experiences through scenery and panorama, smells and sounds as well as the feel of spray and of the wind. Signifying an appreciation of its expanse and openness, Tom noted in bodily terms, "I feel claustrophobic away from the sea". Aesthetics therefore can be seen to be a set of bodily processes, but also culturally situated through different types of landscape interaction. These interactions are partly grounded in gender and class. Men not only place value on hard, physical work that holds purchase within the occupational legacy of the area, but also engage with landscape through masculine leisure activities and sensory experiences that also connect with that domain. This contributes to a local distinctiveness through a topography and legacy that place emphasis on landscape both as an assemblage of physical and structural features and as affective, interrelational and culturally produced.

Landscape, belonging and exclusivity

As Darby (2000) points out, in an active sense landscape 'does' as an instrument of power and cultural practice through which identities are formed. As such, landscape is deeply embedded in relations of power and knowledge, with our understanding and experience of landscape and its imagery 'socially grounded in historically specific notions of exclusion and inclusion' (Darby, 2000: 9). The local distinctiveness of Hastings, with its particular heritage and 'psychic economy', can form the basis of a network of inclusions and exclusions that speak to landscape's access, trespass and belonging, where claims to authenticity, as cited earlier, can be used, in Zukin's terms, as a lever of cultural power. This casts particular groups as outsiders. Incoming European workers, the middle classes arriving from London and those resettled from local boroughs into HMOs in the large Victorian houses that dominate the area were seen to have eroded a strong sense of working -class community ("it was more close-knit … it isn't what it used to be"). Another referred in nostalgic terms to an imagined past, with resentment expressed towards newcomers, presented as having little understanding of heritage and family history, and disrupting long-standing behaviours and norms. "The Old Town is full of weed for my whole life, you know, old fisherman smoking weed. Now those DFLs [down from Londoners] come down and they don't like it outside their posh bloody three-quarters of a million houses" (Pete, painter and decorator).

Notions of belonging were, therefore, tied up with knowledge and experience of the locality, including community practices that were specific to the area and often outside of the mainstream. The claiming of heritage and of the aesthetic were also grounded in residential longevity. As noted earlier, most participants had been brought up in Hastings, with the remainder having moved to the area as children. Resonant with Zukin's (2009) arguments around authenticity, this was often presented as a mark of distinction (one participant claimed, with some pride, "[I'm] Hastings born and bred"). Here, Cresswell (2015) refers to a 'competitive localism' that is exclusive of 'outsiders' and based on origins in the area.

As Burdsey (2011) points out, exclusive, racialized notions of belonging are characteristic of many coastal towns. In our study, whiteness was expressed through common feelings of loss, namely, of a dominant and traditional white, working-class community. In Darby's (2000) terms, landscape can act as an 'abstract repository' of projections, rooted in pre-existing ideological systems that can draw on memory in productions of belonging (Emery, 2018). Ron commented regretfully, "years ago you'd walk round the shopping centre and you'd know everybody ... now you hardly ever hear an English voice". Whiteness as a seemingly threatened condition is therefore inflected with class as long-term residents are positioned against newcomers and where migrants are often presented as contributing to parts of Hastings being 'run down'. Resonant with McDowell and Harris's (2019) Hastings-based study of young men, and as we have noted elsewhere (Simpson et al, 2022), whiteness matters, but 'in a different way' in that it is largely associated with a legacy of belonging based on residential longevity and origins in the area.

Conclusion

With a focus on affective dispositions and experiences and drawing on the broader notion of landscape as both an assemblage of physical/material features and as interrelational and culturally produced, this chapter has explored gendered understandings of landscape and how landscape reflects and engenders affective experiences of belonging. Our context of Hastings, a mid-sized seaside town, reflects its status as one of the most deprived areas in the country, with low relative income levels, above average rates of unemployment (Corfe, 2017) and a geography that contributes potentially to a heavy reliance on manual work.

In creating a cultural identity and a sense of belonging, the physical, socio/cultural and sensory elements of landscape are seen to combine to create a set of 'affective dispositions'. In this respect, the physical features of the town in terms of its coastal location, geographical marginality and inaccessibility are rendered meaningful, reflecting and engendering affects, perceptions and 'ways of seeing' the world. A gendered, affective disposition is manifest in this context through masculine values, including a local dependence and cultural emphasis on manual work, grounded in a localized male employment stereotype that is linked to the geography of the region and a distinctive occupational legacy associated with The Stade. This is accompanied by an attachment to place through the sensory experiences of living near the sea and by masculine recreational activities that are culturally grounded in the area. Landscape is accordingly inhabited and claimed through a classed and gendered 'authenticity' (Zukin, 2009). This generates a collective, masculine identity that crosses occupational boundaries within manual work and where claims to belonging are based on class-based and racialized notions of landscape appreciation and residential longevity. Landscape can accordingly be seen as a material and cultural practice through which class, gender and 'whiteness' are experienced and expressed and through which belonging is constructed and maintained.

We have focused here on some of the affective processes whereby men in our study come to feel included in and identify with a socioterritorial collective, and have highlighted how belonging is based, in both individual and collective terms, on affinity and a strong desire for attachment. Resonant with theories of affect, with its emphasis on relationality, transmissibility and bodily capacity, we can see belonging as an active outcome of encounters with landscape and with other bodies (Deleuze, 1998; Thrift, 2004). As such, it comprises a relational attachment to our environment, born out of interaction and connection so that the affective dimensions of belonging circulate between bodies, shaping social relations and becoming a collective sense. Affective experiences of quietly fishing with a companion, sharing work information in a pub and meeting friends at the angling club contribute to the production of shared values, perceptions, moods and 'states of being' (Hemmings, 2005) – affective dispositions that are internalized within the body. At the same time, as Probyn (1996) suggests, belonging is fragile, precarious and incomplete, threatened by changes in the environment in which people live. In our context, while belonging is evoked through shared values and experiences and through emotional attachment to place, it is at the same time disrupted by incoming residents, manifest in narratives of loss and perceptions of landscape's 'downgrading'.

Bodies, therefore, are implicated in the interpersonal dimensions of community and connection as well as in the internalization of affective dispositions ("it's in our genes"). These are accompanied by the sensorially mediated affective relations of landscape, evoking subjective experiences through sights, touch, smells, tastes and sounds that appeal to our senses, underpinning a place attachment. Our chapter has also highlighted the embodied nature of 'authenticity' (Zukin, 2009) in that claims to landscape and to cultural power are located in the body through politics of inclusion and exclusion, casting some bodies as outsiders. Landscape therefore has material effects on people's lives, feelings and bodily experiences. It is, in a recursive sense, both gendered and gendering in that values and practices impose particular meanings on landscape, while at the same time landscape offers resources for constructions of gendered identity. Both are implicated in the affective experiences of belonging through spatial attachment and sensorial experiences as well as through embodied place making that shapes collective (gendered, classed, raced) identities.

Acknowledgements

This work was supported by the Leverhulme Trust through their Emeritus fellowship scheme: EM-2017-058/7.

References

Acott, T. and Urquhart, J. (2014) Sense of Place and Socio-cultural Values in Fishing Communities along the English Channel. In: J. Urquhart, T. G. Acott, D. Symes and M. Zhao (eds) *Social Issues in Sustainable Fisheries Management*. Springer, pp 257–277.

Anderson, B. (2006) Becoming and Being Hopeful: Towards a Theory of Affect, *Environment and Planning D: Society and Space*: 24: 733–752.

Bender, B. (2002) Time and Landscape. *Current Anthropology*, 43: 103–112.

Bender, B. (2006) Place and Landscape. In: C. Tilley, W. Keane, S. Kuechler, M. Rowlands and P. Spyer (eds) *Handbook of Material Culture*. London: SAGE, pp 303–314.

Burdsey, D. (2011) Strangers on the Shore? Racialized Representation, Identity and In/visibilities of Whiteness at the English Seaside. *Cultural Sociology*, 5(4): 537–552.

Corfe, S. (2017) *Living on the Edge: Britain's Coastal Communities*. Available at: www.smf.co.uk/wp-content/uploads/2017/09/Living-on-the-edge. pdf (accessed 28 February 2019).

Cresswell, T. (2015) *Place: An Introduction*. Chichester: John Wiley and Sons.

Darby, W. (2000) *Landscape and Identity: Geographies of Nation and Class in England*, Oxford: Berg.

Deleuze, G. (1998) *Spinoza. Practical Philosophy*. San Francisco, CA: City Lights Books.

Department for Communities and Local Government (2015) *The English Indices of Deprivation*. National Statistics. Available at: www.gov.uk/gov ernment/statistics/english-indicesof-deprivation-2015 (accessed 29 February 2019).

Emery, J. (2018) Belonging, Memory and History in the North Nottinghamshire Coalfield. *Journal of Historical Geography*, 59: 77–89.

Gregg, M. and Seigworth, G. J. (eds) (2010) *The Affect Theory Reader*. Durham, NC: Duke University Press.

Hastings.gov.uk (2019) Social Housing. Available at: www.hastings.gov.uk/ housing/ social_housing/ (accessed 7 April 2019).

Hemmings, C. (2005) Invoking Affect: Cultural Theory and the Ontological Turn, *Cultural Studies,* 19(5): 548–567.

Jamie, K. (2008) A Lone Enraptured Male. *London Review of Books*, 30(5).

Kraack, A. and Kenway, J. (2002) Place, Time and Stigmatised Youthful Identities: Bad Boys in Paradise. *Journal of Rural Studies*, 18(2): 145–155.

MacFarlane, R. (2007) *The Wild Places*. London: Granta.

Massey, D. (2006) Landscape as a Provocation: Reflections on Moving Mountains. *Journal of Material Culture*, 11(1–2): 33–48.

Massumi, B. (2002) *Parables for the Virtual: Movement, Affect, Sensation*. Durham, NC: Duke University Press.

McDowell, L. (1999) *Gender, Identity and Place: Understanding Feminist Geographies*. Cambridge: Polity Press.

McDowell, L. and Bonner-Thompson, C. (2020) The Other Side of Coastal Towns: Young Men's Precarious Lives on the Margins of England. *Environment and Planning A: Economy and Space*, 52(5): 916–932.

McDowell, L. and Harris, A. (2019) Unruly Bodies and Dangerous Spaces: Masculinity and the Geography of 'Dreadful Enclosures'. *Urban Studies*, 56(2): 419–433.

McDowell, L., Bonner-Thompson, C. and Harris, A. (2020) On the Margins: Young Men's Mundane Experiences of Austerity in English Coastal Towns. *Social and Cultural Geography*, 1–18.

Moore, R. (2019) Is it the End of the Pier for Hastings? *The Observer*, 24th March

Nayak, A. (2003) Last of the 'Real Geordies'? White Masculinities and the Subcultural Response to Deindustrialisation. *Environment and Planning D: Society and Space*, 21(1): 7–25.

Pile, S. (2010) Emotions and Affect in Recent Human Geography. *Transactions of the Institute of British Geographers* 35(1): 5–20.

Probyn, E. (1996) *Outside Belongings*. London: Routledge.

Savage, M., Bagnall, G. and Longhurst B. (2005) *Globalisation and Belonging*. London: Sage.

Sedgwick, E. (2003) *Touching, Feeling: Affect, Pedagogy, Performativity*. Durham, NC: Duke University Press.

Simpson, R., Morgan, R., Lewis, P. and Rumens, N. (2022) Landscape and Work: 'Placing' the Experiences of Male Manual Workers in a UK Seaside Town. *Sociology*, 56(5): 839–858.

Skeggs, B. (2004) *Class, Self and Culture*. London: Routledge.

Thrift, N. (2004) Intensities of Feeling: Towards a Spatial Politics of Affect. *Geografiska Annaler: Series B, Human Geography*, 86(1): 57–78.

Wenham, A. (2020) 'Wish you were here?' Geographies of Exclusion: Young People, Coastal Towns and Marginality. *Journal of Youth Studies*, 23(1): 44–60

Wylie, J. (2009) Landscape, Absence and the Geographies of Love. *Transactions of the Institute of British Geographers*, 34(3): 275–289.

Zukin, S. (1991) *Landscapes of Power: From Detroit to Disney World*, Berkeley: University of California Press.

Zukin, S. (2009) *Naked City: The Death and Life of Authentic Urban Places*. Oxford: Oxford University Press.

Conclusion: Gender, Place and Affect

Ruth Simpson and Alex Simpson

Introduction

We have positioned this edited book at the intersection of gender, place and affect to gain a deeper understanding of spatial and gender-based inequalities. In focusing on how place and gender are intimately constructed and experienced through the circulation and transmission of affect, chapters in this volume, not necessarily discussed in sequence, have explored the significance of affect in shaping our sense of place, identity and belonging and offered insight into the dynamic interplay between social location, the transpersonal processes of affect and gender. Through a lens that is sensitive to the cross-cutting of differences, we have included in our broad category of gender: cisgender, trans, gender queer and non-binary identities, while paying attention to other categories of difference, for example race, class, sexuality and/or whiteness, in terms of how the affective processes of place are shaped, encountered and experienced. With a focus on place as both material and processual, and with an emphasis on gender as *spatial*, *relational* and *embodied* – a corporeal and affective process of becoming that is inseparable from context – we have explored some of the affective and spatial dimensions of advantage and disadvantage.

Attachment, disruption and belonging

The three interrelated, organizing sections have highlighted themes and sub-themes that relate to key affective-geographical formations of *attachment*, *disruption* and *belonging*. In terms of the first, chapters in this section have focused on attachment in the context of work. As we pointed out in our Introduction, workplaces are deeply emotional spaces that involve, among other things, relations, connections, hierarchies and experiences of marginalization. Affect can be seen to 'condition' social relations and be an integral part of organizing, with chapters in this section showing the significance of affective notions of attachment to (and detachment from) norms, logics, objects and people in terms of how men and women negotiate relationships within workplaces, spaces and settings. With a focus on male leadership behaviours in the City of London, Patricia Lewis (Chapter 2) accordingly places emphasis on the significance of postfeminism as an 'affective attachment' in terms of how contemporary leadership is performed, highlighting a distancing on the part of men from individualistic norms and values of masculinity; Darren T. Baker (Chapter 5) focuses on the relationship between space, objects and subject formation and points to how attachment to transitional objects (a child's drawing displayed on a work desk) shapes women's affective lives and work-based identities; Rajeshwari Chennangodu and George Kandathil (Chapter 6) show in the context of a café space how women align with norms and values of the market and connect to those seen to embody market logics in their quest to be seen as a 'better woman worker', while Nick Rumens (Chapter 1) describes his affective distancing and disassociation from a domestic drying rack occasionally positioned, disturbingly, in a space used for his work as a creative artist.

Chapters also speak in different ways to issues around *disruption and disturbances*, the organizing theme of our second section. This theme reflects the unexpected and an unsettling lack of predictability inherent in affective relations, as well as the potential for affect to disrupt the status quo (Sedgwick, 2003). Here, Melissa Tyler, in Chapter 7, highlights how Soho, as an 'affectively queer constellation', disrupts a hegemonic gender regime organized around heteronormatively and a hypermasculine consumption of sex. As she shows, these contestations and struggles are built into Soho's fabric, opening up possibilities for alternative ways of being. Disruption and challenge are themes in Alison Hirst and Christina Schwabenland's Chapter 8 as they demonstrate how, in the confines, insecurities and confusions of Palestinian refugee camps, women create liminal spaces (a kitchen, a roof-top garden) to challenge patriarchal norms and gain a sense of empowerment. The disorders and chaos of homelessness, richly illustrated by Evgeniia Kuziner (Chapter 10) also speak to the significance of 'thick places' (Casey, 2001) that are steeped in meanings and of affective practices as women seek to create the comfort and order of 'home' out of the insecurities and dangers that surround them.

Inherent too in many of these accounts is a focus on *belonging*, a theme we foreground in our third section. As a key dynamic within the affective-geographic experience, belonging relates to our physical involvement with our environment and our embodied 'being in the world' (Emery, 2018). As the chapters in this section show, belonging can be evoked and transmitted through personal and collective memories, emotional attachment to place and collectively shared histories and experiences as well as through boundary making based on manifestations of inclusion and 'non-belonging'. Here, Ruth Simpson and Rachel Morgan (Chapter 14) show in the context of a seaside town how the characteristics of landscape, including its specific historical legacy, are rendered meaningful by white, male manual workers generating affective belonging in terms of the ways in which the landscape is both inhabited and claimed. Drawing on Zukin (1991), they refer to authenticity as the basis for such claims through 'competitive localism' (Creswell, 2015), where a right and affiliation to landscape are mobilized that are exclusive of outsiders. In a different context, Nicholas Hill and his co-authors (Chapter 13) reveal the 'unsettling' of a dominant metronormativity in a small Australian town and how queer lives can be characterized by connection and belonging as young men and women navigate and find ways to live affirmatively in the regions. While, in Chapter 3, Jessica Horne examines the atmospheric relationship between objects and space which shapes the 'angle of arrival' to create a sense of attachment or belonging in a museum setting, in a similar vein, Nyk Robertson's Chapter 12 illustrates the significance of liminality for experiences of belonging where, by breaking free of prescribed gender scripts, non-binary bodies are able to move into spaces otherwise denied to them and feel at home. As Ruth Simpson and Rachel Morgan argue in Chapter 14, drawing on Probyn's (1996) work, belonging is an ongoing, uncertain affective process that is liable to contestation and challenge. Belonging therefore has to be constantly renegotiated in the 'ebbs and flows' of places and of place making.

Taken together, the different contexts outlined in the chapters and the cross-cutting themes of attachment, disruption and belonging reveal the interplay between affective relations and organizational place and space and how they both engender and reflect gendered hierarchies. As intimated earlier, this can be seen in the masculine confidence and power radiated by the architecture of the City of London (Chapter 2) that 'presses' on those who work and move within it as well as the role of domestic artefacts in 'feminizing' space and undermining the status afforded to artistic work (Chapter 1). These central themes accordingly capture the multiple range of differentiated spatial and gendered relationships and the dynamic processes of encounter, re-encounter and sense making which, as we pointed out in the Introduction, help to generate pre-reflexive feelings such as belonging, attachment, community, safety as well as insecurity and fear.

Relationality, transmissibility and bodily capacity

Themes that run *across* the chapters relate to the significance of affect's *relationality, transmissibility* and *bodily capacity* (Deleuze, 1998; Thrift, 2004) for understanding the dynamic interplay between social location, transpersonal processes of affect and gender. As we saw in the previous section, discussions highlight our *relational attachment* to the 'lived, felt' environment in which we live and work, generated through interaction and connection with humans and non-humans, people, places, objects and artefacts, privileging fluidity, proximity and intimacy. Rajeshwari Chennangodu and George Kandathil (Chapter 6), drawing on Deleuze and Guattari, put connectivity at the centre of their analysis of a women-only café space, seeing it as a 'dynamic assemblage' of boundaries, relationships and structures, constantly evolving and 'becoming' in terms of how they are shaped and held together. For Nick Rumens, as we have seen in Chapter 1, a relational attachment to the artefacts of creative artistry (the deployment and display of brushes, easel, pots and paints) helps to construct a personal, masculinized workspace and a valued, professional identity, while the intrusive presence of the domestic drying rack interferes with that construction, generating a 'state of being' that is not conducive to painting.

Affective experiences of place circulate between bodies and things, shaping social relations to become, through *transmissibility and contagion*, a collective sense. This contributes to the production of shared values, perceptions, moods and 'states of being' (Hemmings, 2005) as affective dispositions that are internalized within the body. Here, Murray Lee (Chapter 9) gives insight into how a geography of safety circulates and is reproduced through non-conscious cues and codes, while Patricia Lewis (Chapter 2) traces in the context of the masculine space of the City of London the circulation and transmission of postfeminism as an affective discursive formation in leadership discourse and practice and how male leaders are interpellated to internalize a postfeminist subjectivity. Similarly, Ruth Simpson and Rachel Morgan show in Chapter 14 how a 'psychic economy' based in part on a classed and gendered legacy of a still present fishing industry circulates to generate collective dispositions and practices oriented towards a cultural emphasis on manual work. With an orientation towards the transpersonal, these dispositions, feeling states and action potentials are located *in and between bodies*, both in their capacity to affect and be affected and as a transmission of force or intensity. This transpersonal capacity occurs through 'force encounters' (Gregg and Seigworth, 2010) between bodies and things, engendering a 'becoming' and dovetailing with situation and context. This is neatly captured by Amin and Thrift (2002), referenced in our Introduction, who see affect as a 'whole body cognition' of understanding, experiencing and responding to the world around us, privileging the body through movement and feeling as the site of intensity.

Thus, we can see that practices of walking, socializing, gardening, painting and preparing food are inherently bodily and sensorial, impacting on the self and the body as well as the spaces we traverse and inhabit. Affective dispositions from these encounters, both individually and collectively, are located within the body in the form of gendered perceptions, values and ways of seeing the world, generative of practice. Corina Sheerin and Alex Simpson (Chapter 4) demonstrate how finance workers walking through the streets of the City of London can internalize the scale and grandeur of the buildings around them into exclusive, gendered constructions of individual and collective entitlement and success. These practices, in turn, impact on the meanings attached to place through, for example, a fast pace of walking and a 'corporate', professional comportment which help appropriate the City as a masculine domain, confirming its status as a space of money-making and purpose. Masculine dispositions are accordingly both 'within the body' and woven into the fabric of the City more widely, transmitted through collectively shared notions (competition, efficiency, confidence) and embodied practice. These affective relations between the body and the material form are also captured in this volume in Alison Hirst and Christina Schwabenland's study (Chapter 8) of a Palestinian refugee camp. Here, in a roof-top garden, women engage in affective, bodily practices of food production, traditionally associated with women, while at the same time repairing, restoring and replanting the garden following its destruction by spates of enemy attacks, transforming themselves into resilient activists through their stoic resistance and stirring between them quiet, positive response. Taken together, this highlights the 'space in between' (Pile, 2010) in the flow of affect, as a relational force, between bodies, between bodies and 'things' as well as between bodies and the spatial environment.

Potential, place and affect

Following Pile (2010) and Thrift (2004), the collected work within this book suggests that, as a composition of 'force encounters' (Gregg and Seigworth, 2010), affect has strong action potential. Lying between individuals and between individuals and place, affect turns on its intensity and capacity and potential to do. This potential is transmitted between bodies, where bodies include an array of human, non-human actors, objects and processes that also interact with and partly constitute the materiality, history and culture of place. However, as Hemmings (2005) deftly explains, questions remain as to the nature and direction of this transformational capacity. Here, Thrift (2004) expresses concerns over the power dynamic whereby corporations and state seek the control of affect to meet their own ends. Others (for example, Massumi, 2002; Sedgwick, 2003) place emphasis on affect's positive, transformative power. As they suggest, imbued with creative possibility and with potential to be unsettling through the unexpected and the singular, affect carries the capacity to exceed the social subjection, suggested by Thrift, cited earlier, and disrupt the status quo. Rather than this dichotomy of positive/ negative, critiqued by Hemmings (2005), our chapters suggest an evolving simultaneity of subjection and emancipation where, in a less ambitious sense, some gender relations may be quietly challenged and hegemonic meanings restructured, while others are further entrenched. As Berlant wryly notes 'shifts in affective dispositions are not equal to changing the world' (Berlant, 2010: 100). In other words, the network of 'force encounters' and the 'affective assemblages' (Wetherell, 2012) of everyday life can elicit micro-outcomes that, at a local level, may partly disrupt and/or go some way to further confirm the dominant order.

Thus, Melissa Tyler (Chapter 7) positions the affective atmosphere of Soho in contradictory terms both as hegemonically masculine and heteronormative *and* as critically queer and multiplicitous, with the former embedded into the materiality and practices of space. As she argues, as an oasis of 'Otherness', the social materiality of Soho is heavily encoded with heteronormative, gendered imagery, *at the same time* as it opens up the possibility for this to be challenged and for queer politics and gender multiplicity to emerge. Equally, Alison Hirst and Christina Schwabenland (Chapter 8) show how, in the context of a refugee camp kitchen where women prepare and sell food, gender norms are both confirmed (in the domestic role of food production) and challenged (as they develop the capacities of businesswomen). As they conclude: 'These improvisations in affective practices have reconfigured the "affective ruts" (Wetherell, 2012: 13) laid down for many decades, not through a spectacular or sensational change but by small, iterative, gentle, changes that build on but do not directly challenge women's traditional roles. They stir positive emotional responses, but quietly.'

This 'quiet challenge' to the dominant order emerges in Nicholas Hill's chapter (Chapter 13) where he and his co-authors describe the 'unsettling' of metronormative assumptions in the regions of Australia. Darren T. Baker (Chapter 5) refers to how desk objects can, in a small way, give women a sense of agency and comfort as they 'live through' the challenges of an office environment. Nyk Robertson (Chapter 12), looking at spaces of belonging for non-binary folx, demonstrates in more conditional terms the potential role of 'possibility scripts' in expanding opportunities for queer bodies. This points to the simultaneity of challenge to and confirmation of the dominant social order as well as more localized, muted, provisional and tentative outcomes than the 'transformation' discourse would suggest.

Place making, atmospheres and intensity

These chapters also to speak to the gendered and affective processes involved in 'place making', where places are seen to be constituted and maintained by social relations of power and exclusion (McDowell, 1999). Affects and practice come together to actively constitute or produce places in ways that are intertwined with gender, whether it is a women-only café, a space for creative artistry in a conservatory, a 'home' in a derelict building or the 'making and re-making' of places such as Soho, Barangaroo and the City of London. As Evgeniia Kuziner explains in Chapter 10, constructions of 'home' as an affective site of domesticity, security and emotional comfort are shaped by traditional understandings and practices of gender relations where women take up a role and subjectivity of 'hostess' and principal carer to the family. In other words, gender shapes how we understand and experience place, which in turn has implications for gendered hierarchies and for gendered subjectivities. This points to both practical and affective dimensions to place making, and here Duff (2010) refers to 'affective atmospheres' as flows of affect. These capture the affective mood of place, the dispositions and agencies as well as the activities enactable within it that elicit an emotive and embodied response. As Alex Simpson and Paul McGuinness point out in Chapter 11, drawing on Bissell (2018), affective atmospheres are perceived and sensed through the body, woven into body discourse to create an often pre-reflexive experience of space. In other words, as Anderson (2009) summarizes, located between experiences and setting, affective atmospheres can be seen as a relation of felt connections that both shape and are shaped by the surrounding environment.

As illustration, Melissa Tyler (Chapter 7) describes how the affective atmosphere of Soho is encoded (for example, as hyper-heteronormative, gender multiplicitous and queer) and how this atmosphere is embedded and enacted through meanings, practices and materialities of its various 'streets, courts and alleyways', as well as through the physical, embodied presence of those who live, work, consume, pass through or otherwise 'hang out' in the area. Nick Rumens (Chapter 1) charts in historical terms the affective atmosphere and the gendering of the home conservatory, associated in its original form with novelty, exploration and the exotic, and how affective processes of place making in this context are interwoven with colonial masculinity, informing practice and behaviour. As he illustrates, drawing on Schiebinger (2007), the exotic and ornamental plants that were housed in conservatories reflect and reproduce imperial discourses of European colonial masculinity as a positive and progressive force. Alex Simpson and Paul McGuinness show in Chapter 11 how the financial architecture (new, glistening glass edifices) and the production and framing of Barangaroo in Sydney's central business district privilege a masculinist atmosphere that reflects and is entwined with the masculinity and exclusivity of finance as a profession, a source of demarcation that imprints on the gendered dynamics of identity formation and belonging.

These examples highlight how feeling states and meanings, together with associated practices, accrue over time, comprising the ongoing production of place as lived, felt and relational embodied experience. Simpson and McGuiness, for example, chart the affective history of Barangaroo, where development is packaged as 'urban regeneration', concealing the erasure of the dockland that preceded it as well as the unceded lands of the Gadigal people. In short, as Duff (2010) notes, drawing on Thrift (2004), place making draws on affective atmospheres as diverse social and affective resources that are layered with erasure and with past meanings, and which require an active engagement, through practice, with the material dimensions of place. The lived, practical and affective experience of place is captured by Casey (2001) in terms of its intensity and depth, highlighting how places are worked into 'thick' meaningful places to which individuals have an affinity. This requires an affective engagement with place, giving opportunity for personal enrichment and a 'deepening of affective experience' (Duff, 2010: 881). In contrast to 'thin' places that do not elicit a strong sense of affiliation, thick places are made in and of affect and practice, where the latter is imbued with affective force, structuring experiences of self and belonging. In this way, practices of gardening and cultivation (weeding, sowing, tending plants) help to construct a roof-top garden in a refugee camp, giving meaning and a sense of belonging to those involved through an 'intensification of the affective pull' of place (Duff, 2010: 882).

While not necessarily drawing on this terminology, our chapters show how 'thick' places are constituted and maintained, and the significance of gender in their production and reproduction. Evgeniia Kuziner (Chapter 10) highlights the dangers and insecurities facing homeless women and how they create domestic spaces of comfort and security (through furnishing, practices of cleanliness, ownership of pets) 'deep within' derelict buildings for greater invisibility and protection from authorities who may seek to evict them. Rajeshwari Chennangodu and George Kandathil chart (Chapter 6) the uncertain and often unstable emergence of a café space and its position as a source of hope and expectation, comprising a 'collective assemblage of empowerment' in women's intense desire to develop themselves and overcome disadvantage. Nyk Robertson (Chapter 12) argues for the potential for queer, liminal spaces to be a source of belonging and community, where intersex bodies can be 'known', while in a similar vein Nicholas Hill and his co-authors (Chapter 13) find affective connections, joyfulness and community for LGBTIQA+ youth in queer-friendly places in the regions of Australia, such as a local café, bookstore and theatre. However, in Murray Lee's Chapter 9 we can see that such places may not always engender positive affects of community, attachment and belonging, and here he refers to how perceptions of safety can negatively attach to places as sites of danger. These places may still be seen as 'thick' in that they are invested, as Casey (2001) suggests, with intense meaning. They carry affective connections that form a 'deep affective experience'. These impinge on the sense of self but, generative of insecurity and fear, are to be repulsed and avoided. Melissa Tyler, in her chapter on Soho, refers here to 'abject spaces' and the coexistence of places that are both 'seductive and repulsive' where the latter are ' "thick" in their affective atmospheres and associations, but which do not necessarily support enriching experiences or nurture close affinities'. As suggested by Murray's chapter, while 'thick' places in a positive sense can create 'zones of belonging' Casey (2001), in other contexts they may carry the capacity for negative feelings of insecurity and abjection, reflective too of gender dynamics and hierarchies in the ways in which such places are constructed and maintained.

Gender, place and affect

In this volume, we have conceptualized gender as intimately connected to the transpersonal dynamics of affect, namely as a virtual and corporeal process of becoming and possibility that is inseparable from context. With a focus on affect's relationality, transmissibility and bodily capacity, we have highlighted key affective processes and the implications for gendered relations and gendered hierarchies in the form of *attachment* to a lived felt environment as well as to norms, practices, people, artefacts and ideas; *disruption and disturbances* through the unexpected, the uncontrollable and challenges to the dominant order; and *belonging* as an individual and collective experience of place, integral to subject formation and to processes of place making. This is to see gender as simultaneously *spatial*, *relational* and *embodied* and to put emphasis on the role of *affective atmosphere*, in terms of how gender is experienced and transmitted in context. A spatial understanding of gender is to recognize that, as an affective process of becoming that comprises an embodied relationship with the world (Massey, 1994), gender is inseparable from context and its mix of social relations. This relational understanding sees gender as embedded in and emerging out of flows of interactions and encounters, highlighting the gendered nature of the 'affective force relations' that, in Gregg and Seigworth's (2010) terms, exist 'in-between'. These force relations are located both within and between bodies as a gendered 'whole body cognition' (Amin and Thrift, 2002), where gendered hierarchies are encountered and 'felt' in the form of circulating, embodied dispositions and where, through action potential, such hierarchies are reproduced or challenged.

The spatial, relational and embodied nature of gender is to highlight that gender relations are not just located within the body in the form of 'feelings', pre-reflexive senses or dispositions, but circulate in context. As we have seen, affective atmospheres of place reproduce and reflect gendered hierarchies, woven 'thickly' into place as well as into discourse, impressing on individual men and women and shaping (and shaped by) the diverse practical and affective processes of place making. As our chapters have sought to demonstrate, the dynamic, spatial processes of these felt connections are integral to understanding the lived experience of gender in place.

References

Amin, A. and Thrift, N. (2002) *Cities: Reimagining the Urban.* Cambridge: Polity Press.

Anderson, B. (2009) Affective Atmospheres. *Emotion, Space and Society,* 2(2): 77–81.

Berlant, L. (2010) Cruel Optimism. In: M. Gregg and S. J. Seigworth (eds) *The Affect Theory Reader.* Durham, NC: Duke University Press, pp 93–117.

Bissell, D. (2018) *Transit Life: How Commuting is Transforming Our Cities.* Cambridge, MA: MIT Press.

Casey, E. (2001) Between Geography and Philosophy: What Does It Mean to Be in the Place-world? *Annals of the Association of American Geographers,* 91: 683–693.

Cresswell, T. (2015) *Place: An Introduction* (Second edition). J. Wiley and Sons.

Deleuze, G. (1998) *Spinoza. Practical Philosophy.* San Francisco, CA: City Lights Books.

Duff, C. (2010) On the Role of Affect and Practice in the Production of Place. *Environment and Planning D: Society and Space,* 28(5): 881–895.

Emery, J. (2018) Belonging, Memory and History in the North Nottinghamshire Coalfield. *Journal of Historical Geography,* 59: 77–89.

Gregg, M. and Seigworth, G. J. (2010) An Inventory of Shimmers. In: M. Gregg and G. J. Seigworth (eds) *The Affect Theory Reader.* Durham, NC: Duke University Press, pp 1–25.

Hemmings, C. (2005) Invoking Affect: Cultural Theory and the Ontological Turn. *Cultural Studies,* 19(5): 548–567.

Massey, D. (1994) *Space, Place and Gender.* Cambridge: Polity Press.

Massumi, B. (2002) *Parables for the Virtual: Movement, Affect, Sensation.* Durham, NC: Duke University Press.

McDowell, L. (1999) *Gender, Identity and Place: Understanding Feminist Geographies.* Cambridge: Polity Press

Pile, S. (2010) Emotions and Affect in Recent Human Geography. *Transactions of the Institute of British Geographers,* 35: 5–20.

Probyn, E. (1996) *Outside Belongings.* London: Routledge

Schiebinger, L. (2007) *Plants and Empire: Colonial bioprospecting in the Atlantic World.* Cambridge, MA: Harvard University Press.

Sedgwick, E. K. (2003) *Touching Feeling: Affect, Pedagogy, Performativity.* Durham, NC: Duke University Press.

Thrift, N. (2004) Intensities of Feeling: Towards a Spatial Geography of Affect. *Geografisko Annaler Series B,* 86: 57–78.

Wetherell, M. (2012) *Affect and Emotion: A New Social Science Understanding.* London: Sage.

Zukin, S. (1991) *Landscapes of Power: From Detroit to Disney World.* Berkeley: University of California Press.

Index

Note: References to figures appear in *italic* type. References to endnotes show both the page number and the note number (150n1).